Ocean Engineering & Oceanography

Volume 10

Series editors

Manhar R. Dhanak, Florida Atlantic University SeaTech, Dania Beach, USA
Nikolas I. Xiros, New Orleans, USA

More information about this series at http://www.springer.com/series/10524

Luc Jaulin · Andrea Caiti
Marc Carreras · Vincent Creuze
Frédéric Plumet · Benoît Zerr
Annick Billon-Coat
Editors

Marine Robotics and Applications

Springer

Editors
Luc Jaulin
LABSTICC
ENSTA Bretagne
Brest
France

Frédéric Plumet
ISIR
University Pierre and Marie Curie
Paris
France

Andrea Caiti
Department of Information Engineering, Centro di Ricerca Enrico Piaggio
University of Pisa
Pisa
Italy

Benoît Zerr
Ocean Sensing and Mapping
ENSTA Bretagne
Brest
France

Annick Billon-Coat
Lab-STICC
ENSTA Bretagne
Brest
France

Marc Carreras
Escola Politècnica Superior (UdG), Edifi, Computer Vision and Robotics Institute
University of Girona
Girona
Spain

Vincent Creuze
Laboratoire d'Informatique, de Robotique et de Microélectronique de Montpellier (LIRMM)
Montpellier
France

ISSN 2194-6396 ISSN 2194-640X (electronic)
Ocean Engineering & Oceanography
ISBN 978-3-030-09985-5 ISBN 978-3-319-70724-2 (eBook)
https://doi.org/10.1007/978-3-319-70724-2

Printed on acid-free paper

This Springer imprint is published by Springer Nature
The registered company is Springer International Publishing AG
The registered company address is: Gewerbestrasse 11, 6330 Cham, Switzerland

Preface

Every 2 years, the MOQESM conference takes place in Brest (France), in the framework of the SeaTech Week Conference. MOQESM is the meeting point of specialists from two fields, namely coastal hydrography and marine robotics. This is a unique opportunity for industrials and academic scientists to present their respective works to each other. It often leads to very rich scientific exchanges during and after the presentations. Indeed, many hydrographic applications can be performed by AUVs (Autonomous Underwater Vehicles), and, reciprocally, many navigation methods can benefit from the advances in hydrography for localization purpose, target identification, or path planning. The conference is also open to more general marine robotic presentations in fields like design and operation. This book gives to the reader an overview of some of the topics addressed during the MOQESM 16 Conference (October 11–12, 2016). This book is more than the proceedings of the conference, as it includes also original papers that have not been presented during MOQESM, but have been selected by the editors. The topics covered by this book include the following:

1. Acoustics: sonar tracks registration
2. Localization and navigation: simultaneous localization and mapping, range only localization, interval analysis for trajectory estimation, and electric sense for navigation
3. Multi-vehicles methods: securing an area with AUVs and performing adaptive sampling with autonomous sailing boats
4. Design: optimization of propulsion systems, and design and control of an underwater vehicle.

If this is not already the case, we hope that this book will convince you of the interest of bringing coastal hydrography and marine robotics together and we hope that it will make you as enthusiastic as we are to participate to next MOQESM editions.

Montpellier, France
May 2017

Vincent Creuze

Contents

Chapter 1
Fast Fourier-Based Block-Matching Algorithm for Sonar Tracks Registration in a Multiresolution Framework

Florian Nicolas, Andreas Arnold-Bos, Isabelle Quidu and Benoît Zerr

Abstract In the underwater mine warfare context, potential threats are usually detected and classified by means of an Automatic Target Recognition (ATR) chain, especially in case of newly surveyed areas. However, if we can rely on a previously acquired sonar track, it is conceivable to directly compare such a track, said as reference, with a more recent one in order to detect seabed changes such as new objects lying on the seabed. To perform this change detection process, the very first step consists in geometrically aligning the reference and the repeated tracks. In this paper, we detail a block-matching approach using masked Fourier cross-correlation as a similarity metric, to carry out a fast elastic registration in a multi resolution framework. To improve the robustness of the algorithm, the resulting vector field is then filtered thanks to the navigation uncertainty, provided by the INS, along with an Inverse Distance Weighting estimate, to get rid of outliers.

1.1 Introduction

In the context of sonar imagery, image registration, consisting in geometrically aligning two or more images, is a crucial step as it is often the very first step to further perform several tasks such as navigation correction [1–3], seabed mosaicking

---

The original version of this chapter was revised: For detailed information please see Erratum. The erratum to this chapter is available at https://doi.org/10.1007/978-3-319-70724-2_10

F. Nicolas (✉) · A. Arnold-Bos
Thales Underwater Systems, Brest, France
e-mail: florian.nicolas@fr.thalesgroup.com; florian.nicolas@ensta-bretagne.org

A. Arnold-Bos
e-mail: andreas.arnold@fr.thalesgroup.com

F. Nicolas · I. Quidu · B. Zerr
Lab-STICC/PRASYS, UMR CNRS 6285, ENSTA Bretagne, Brest, France
e-mail: isabelle.quidu@ensta-bretagne.fr

B. Zerr
e-mail: benoit.zerr@ensta-bretagne.fr

L. Jaulin et al. (eds.), *Marine Robotics and Applications*, Ocean Engineering & Oceanography 10, https://doi.org/10.1007/978-3-319-70724-2_1

[4–6] or also to carry out change detection [7–10]. Such an alignment can therefore be achieved through various methods. We usually split these methods into two groups, the symbolic ones (features-based) [1–5, 7, 11–14] and the iconic ones (intensity-based) [6, 8, 9, 15–20]. While the former try to extract and match features such as objects lying on the seabed, salient points or homogenous areas, the latter directly work with the pixels intensity through various similarity (resp. dissimilarity) measures to maximize (resp.minimize), to find the optimal transformation.

As we are looking for a method which does not rely on any features detection and description step, we focus on intensity-based methods. Indeed, we want such an algorithm to apply whatever the seabed type (presence of man-made objects or not, presence of sand ripples or not, textured seabed). A key point of a block matching algorithm is the choice of the similarity metric. The most famous one is undoubtedly the mutual information [21] which has massively been used by the medical image processing community [22] but also in the underwater field [17, 18]. However, according to Vandrish [3], such a metric can perform poorly compared to other similarity metrics such as the phase correlation. In addition, in [23], we demonstrated that a zero normalized cross-correlation (ZNCC) can be sufficient to perform registration. Moreover, the ZNCC, computed in the Fourier domain, retrieves the optimum through an unique iteration, while other metrics will have to look for the maximum in the entire search space, requiring much more iterations. As our algorithm should be designed to operate in real-time, such an advantage cannot be neglected. Although it could be conceivable to search for such a maximum through an optimization process, we can not insure to converge towards the global optimum. Moreover, the masked version of the Fourier NCC proposed by Padfield [24] is of interest in our case, allowing us to only take into account some parts of sonar tracks during registration.

In Sect. 1.2, we emphasize the key steps of the proposed block-matching algorithm which results on real high-resolution sonar tracks are provided in Sect. 1.3.

1.2 Non-rigid Registration Algorithm

1.2.1 Masked Normalized Cross-Correlation

As it has previously been shown in a rigid context [23], the zero normalized cross-correlation (ZNCC) is suitable to perform sonar tracks registration. The ZNCC between two images f_1 and f_2 can be expressed, in the spatial domain, as:

$$ZNCC(u,v) = \frac{\sum\limits_{(x,y)\in\cap_{u,v}} (f_1(x,y) - \overline{f_1})(f_2(x-u,y-v) - \overline{f_2^{(u,v)}})}{\sqrt{\sum\limits_{(x,y)\in\cap_{u,v}} (f_1(x,y) - \overline{f_1})^2}\sqrt{\sum\limits_{(x,y)\in\cap_{u,v}} (f_2(x-u,y-v) - \overline{f_2^{(u,v)}})^2}} \tag{1.1}$$

where $\cap_{u,v} = \{(x, y) \in f_1 \cap f_2^{(u,v)}\}$ where $f_2^{(u,v)}$ is a shifted version of f_2 with shifts u, v, and $\bar{f_i}$ corresponds to the average of f_i on this same interval.

In addition, it is possible to insert binary masks (m_i for image f_i) in the ZNCC Fourier formulation (1.2) in order to discard certain regions from both images [24] as they would affect the registration results adversely. This is known as the Masked Normalized Cross-Correlation (MNCC). Thus, in our underwater context, such an ability is welcome for various reasons. Firstly, as we deal with geo-referenced sonar tracks, it allows us to register blocks located near the sonar track borders while only taking into account pixels associated with sonar data. Secondly, as reference and repeated tracks can differ from each other due to differences in their point of view and/or grazing angle, it could sometimes be interesting to remove shadows from the mask to achieve registration. However, we do not need to retrieve a scale factor and a rotation factor. Indeed, we assume that these global parameters can be modeled through our block-matching formulation.

$$
\begin{aligned}
MNCC &= \frac{\mathcal{F}^{-1}(F_2 \odot F_1^*) - \frac{\mathcal{F}^{-1}(F_2 \odot M_1^*) \odot \mathcal{F}^{-1}(M_2 \odot F_1^*)}{\mathcal{F}^{-1}(M_2 \odot M_1^*)}}{\sqrt{\mathcal{F}^{-1}(\mathcal{F}(f_2 \odot f_2) \odot M_1^*) - \frac{(\mathcal{F}^{-1}(F_2 \odot M_1^*))^2}{\mathcal{F}^{-1}(M_2 \odot M_1^*)}} \sqrt{\mathcal{F}^{-1}(M_2 \odot \mathcal{F}(f_1' \odot f_1')) - \frac{(\mathcal{F}^{-1}(M_2 \odot F_1^*))^2}{\mathcal{F}^{-1}(M_2 \odot M_1^*)}}} \\
m_1' &= rot(m_1, 180^\circ) \\
f_1' &= rot(f_1, 180^\circ) \\
M_2 &= \mathcal{F}(m_2) \\
M_1^* &= \mathcal{F}(m_1') \\
&\odot \text{ is the element-wise product operator}
\end{aligned}
\tag{1.2}
$$

1.2.2 *Block-Matching Algorithm*

As demonstrated in [23], a rigid body transformation can not be chosen to precisely register two sonar tracks. Indeed, sonar data projected in the Earth frame are prone to error due to carrier uncertainty navigation. To overcome such an issue, as detailed in Fig. 1.1, our algorithm first consists in dividing the reference track into several blocks at the first available resolution level by means of a regular grid. The size of such blocks should be small enough to be able to model local deformations but large enough to contain sufficient information about the seabed. Then, a MNCC is computed between such a block from the reference image and another larger one from the repeated track. Figure 1.2 gives two examples of the MNCC between two blocks. Once all blocks have been processed, the yielded vector field must undergo a two-steps filtering as it will be explained in the next Sects. (1.2.3 and 1.2.4). Such a filtered vector field is then used to initialize the registration at the next finer level, allowing to decrease the search space and thus decreasing the computational cost. Finally, the vector field obtained at the finest resolution is used to perform the repeated track warping (Sect. 1.2.6).

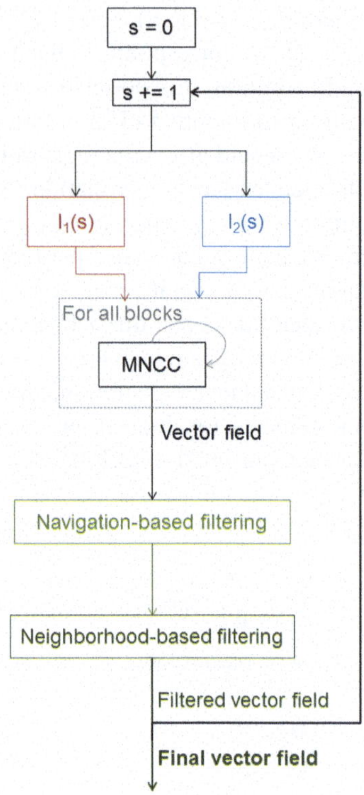

Fig. 1.1 Overview of our multi-resolution block-matching algorithm

1.2.3 Navigation-Based Filtering

As the peak location provided by the MNCC does not necessarily correspond to a physically reliable displacement, we rely on the latitude and longitude uncertainties provided by the Inertial Navigation System (INS), for each transmitted ping, to get rid of such outliers. Such outliers can for example be due to repetitive patterns such as linear trawling net marks or blocks located within homogeneous areas where the correlation peak can be sensitive due to the lack of distinctive features.

Thus, let p_i^{ref} (resp. p_i^{rep}) be the observation of the ith true contact p_i (unobserved) acquired during the reference (resp. repeated) pass:

$$\begin{aligned} p_i^{ref} &\sim \mathcal{N}(p_i, \Sigma_i^{ref}) \\ p_i^{rep} &\sim \mathcal{N}(p_i, \Sigma_i^{rep}) \end{aligned} \tag{1.3}$$

Σ_i^{ref} (resp. Σ_i^{rep}) is the reference (resp. repeated) diagonal covariance matrice.

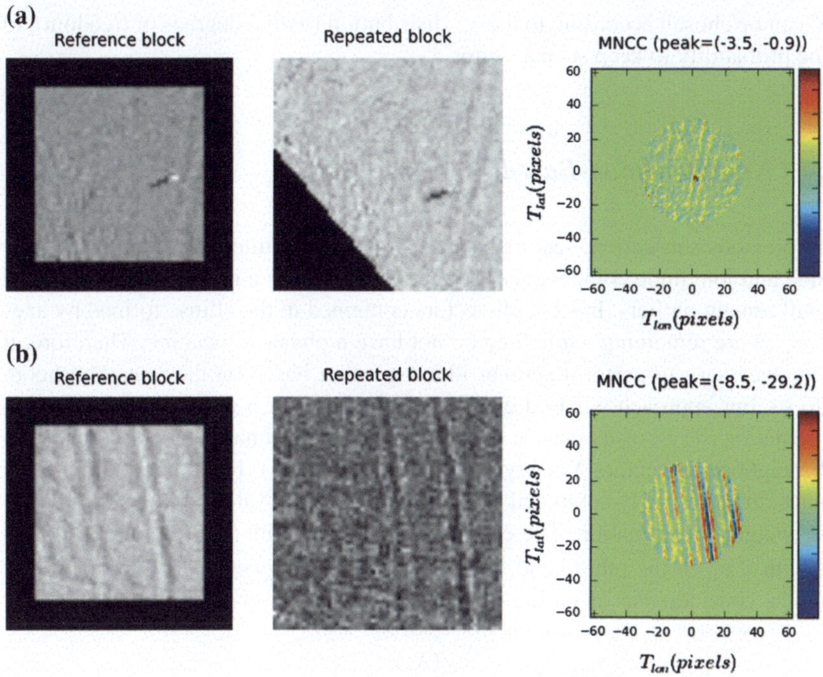

Fig. 1.2 Examples of masked normalized cross-correlation computed at the coarsest resolution level. **a** MNCC yielding the expected displacement between both blocks owing to the mine-like contact. **b** MNCC providing an erroneous displacement due to the linear trawling net mark raising an ambiguity

As p_i^{ref} and p_i^{rep} are considered independent, the distribution of $d_i = p_i^{ref} - p_i^{rep}$ can be written as:

$$\begin{aligned} d_i &\sim \mathcal{N}(0, \Sigma_i^{ref} + \Sigma_i^{rep}) \\ &\sim \mathcal{N}(0, \Sigma_i) \end{aligned} \tag{1.4}$$

with:

$$\begin{aligned} \Sigma_i &= \begin{pmatrix} (\sigma_{i,lat}^{ref})^2 + (\sigma_{i,lat}^{rep})^2 & 0 \\ 0 & (\sigma_{i,lon}^{ref})^2 + (\sigma_{i,lon}^{rep})^2 \end{pmatrix} \\ &= \begin{pmatrix} \sigma_{i,lat}^2 & 0 \\ 0 & \sigma_{i,lon}^2 \end{pmatrix} \end{aligned} \tag{1.5}$$

A registration vector $v_i \begin{pmatrix} \Delta_{i,lat} \\ \Delta_{i,lon} \end{pmatrix}$ will thus be kept if and only if, it is contained in the following uncertainty ellipsis:

$$\left(\frac{\Delta_{i,lat}}{\sigma_{i,lat}}\right)^2 + \left(\frac{\Delta_{i,lon}}{\sigma_{i,lon}}\right)^2 \leq K \tag{1.6}$$

K can be chosen according to the χ^2 distribution (with 2 degrees of freedom), to set the probability to keep such a vector.

1.2.4 Neighborhood-Based Filtering

In the previous subsection, vectors have separately been filtered in examining their latitude and longitude components. At the output of such a norm-based filtering, it can still remain outliers. Indeed, all vectors contained in the ellipse formed by axes $(\sigma_{lat}, \sigma_{lon})$ are remaining while they do not have a physical meaning. Therefore, it is now necessary to perform another filtering rather based on their neighborhood. To do so, our approach is based on the computation, for a given vector X (origin x, orientation $\theta(x)$), of its inverse distance weighting estimate $\widehat{\theta}(x)$ (1.7) thanks to its N neighboring vectors X_i (origin x_i, orientation $\theta(X_i)$) (1.7). In addition to the distance, the previously computed cross-correlation peak value $\gamma(X_i)$ is introduced in the weights computation. The comparison of $\theta(X)$ with $\widehat{\theta}(X)$ in one hand, and $\widehat{\|X\|}$ with $\|X\|$ in the other hand, thus allows to decide wether the current vector should be considered as an outlier (1.8). T_θ and $T_{\|.\|}$ are user-defined thresholds.

$$\omega(X, X_i) = \frac{\gamma(X_i)}{d(x, x_i)^p}$$

$$\widehat{\theta}(X) = \frac{\sum_{i=1}^{N} \theta(X_i)\omega(X, X_i)}{\sum_{i=1}^{N} \omega(X, X_i)} \tag{1.7}$$

$$\widehat{\|X\|} = \frac{\sum_{i=1}^{N} \|X_i\|\omega(X, X_i)}{\sum_{i=1}^{N} \omega(X, X_i)}$$

$$\|\widehat{\theta}(X) - \theta(X)\| \leq T_\theta$$

$$\frac{1}{T_{\|.\|}} \leq \frac{\widehat{\|X\|}}{\|X\|} \leq T_{\|.\|} \tag{1.8}$$

The vector field variation should obviously be considered as slow for our approach to be valid.

1.2.5 Multi-resolution Framework

The multi-resolution framework consists in initializing the next level thanks to the vector field computed at the previous one (Fig. 1.3). Thus, this allows to reduce the search space along the resolution levels (denoted by s) i.e. constraining the vector field while speeding up the algorithm.

As some vectors will be dismissed once filtering steps are done, it is possible to generate new vectors at the next scale by duplicating inlying vectors in their neighborhood in order not to run out of vectors.

Regarding the blocks size, a compromise has to be found. Indeed, let S_{ref} (resp. S_{rep}) be the size of a block from the reference track (resp. repeated track). S_{ref} should be large enough to contain enough information to locally describe the seabed but narrow enough to be able to model local deformations. However, S_{rep}, which is relatively large at the very first scale, decreases along the resolution levels as the correlation peak is more and more accurately localized.

1.2.6 Warping Functions

Once the final vector field has been obtained, the next step consists in computing a transformation for every pixel belonging to the repeated track. Several transformation functions have been considered such as the radial basis functions whose general form is given in (1.9). The radial basis function $R(.)$ only depends on the distance between the ith data point and the point x ($x \in \mathbb{R}^2$ in the case of 2D image registration) to interpolate.

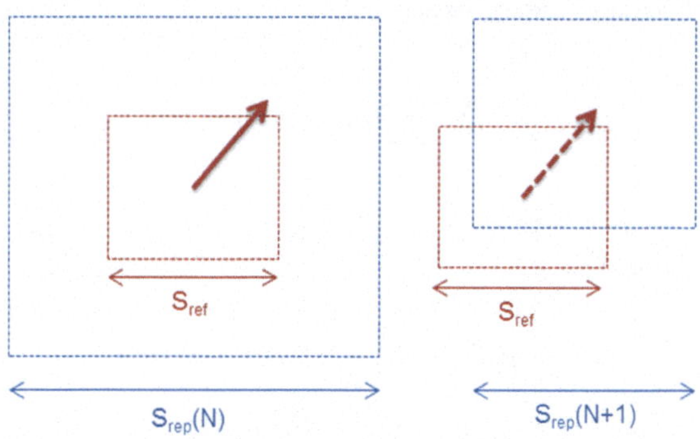

Fig. 1.3 (*from left to right*). At the coarsest level, for both blocks from the reference (red dotted square) and repeated (blue dotted square) tracks, the MNCC provides a displacement vector (red arrow). Thus, at the next level, the location of the block from the repeated track is initialized by means of the previously computed vector. This allows to decrease the search space along the resolution levels to speed up the registration process

$$f(x) = \sum_{i=1}^{n} \alpha_i \underbrace{R(d_i(x))}_{\text{Radial basis function}} + \underbrace{p_m(\boldsymbol{x})}_{\text{polynomial of degree m}} \tag{1.9}$$

The most famous RBF function is probably the Thin Plate Spline (TPS) (1.10) which has extensively been used in image registration.

$$R(d_i) = d_i^2 log(d_i) \tag{1.10}$$

However, in our case, although the center of blocks are arranged according to a regular grid, our algorithm may finally yield a sparse vector field whose density significantly varies in the space (vectors can thus all be located in a very small area of the sonar track). Thus, we often deal with extrapolation rather than interpolation, and, as the TPS have no local support property, it often fails to provide a reliable field.

Moreover, as we target real time processing, the interpolation complexity is a key factor. Indeed, while the TPS complexity is $O(nN)$ [25], n and N respectively being the number of data points and the number of points we want to interpolate, we prefer using faster methods such as triangulation-based ones or B-splines. As their name suggests, triangulation-based methods consists in triangulating the data points through Delaunay triangulation and then perform piecewise interpolation in each triangle by means of linear functions or cubic ones. A con of triangulation-based methods comes from the fact that they are only able to perform the interpolation inside the convex hull. To overcome this, we use the Shepard's method to estimate the vector field on sonar track borders and thus extend the convex hull.

Regarding B-splines methods, they have broadly been used in non-rigid and optimization-based medical image registration [26, 27], as they are locally supported (i.e. it only relies on some neighbors to perform the interpolation). Given a 2D control lattice ϕ, they can be expressed as

$$f(x, y) = \sum_{k=0}^{3} \sum_{l=0}^{3} B_k(s) B_l(t) \phi_{i+k, j+l}, \tag{1.11}$$

where $i = \lfloor x \rfloor - 1$, $j = \lfloor y \rfloor - 1$, $s = \lfloor x \rfloor - x$, $t = \lfloor y \rfloor - y$. B_k and B_l are B-spline basis functions defined as

$$\begin{aligned} B_0(t) &= \frac{(1 - t^3)}{6}, \\ B_1(t) &= \frac{(3t^3 - 6t^2 + 4)}{6}, \\ B_2(t) &= \frac{(3t^3 - 6t^2 + 4)}{6}, \\ B_3(t) &= \frac{t^3}{6}. \end{aligned} \tag{1.12}$$

In Sect. 1.3.2, multilevel B-splines [28] and a triangulation-based approach [29] are compared.

1.3 Experimental Results

1.3.1 Dataset

To evaluate our registration algorithm, we have two high-resolution synthetic aperture sonar tracks acquired offshore Brittany coasts, near Brest, on September 2015. Such data have been acquired with the SAMDIS sonar sensor manufactured by Thales. While this sensor is able to perform multi-aspect acquisition, we only rely on the broadside aspect to perform our registration. The carrier is also equipped with an iXBlue PHINS inertial navigation system whose data are fused with a Teledyne RDI Workhorse 300 data velocity log (DVL). First and foremost, we project both sonar tracks in the Earth frame before storing them at different resolutions through a hierarchical data format.

We have chosen three different resolution levels, ranging from 0.5 m to 0.12 m, to benefit from the multi-resolution framework. A given resolution level of a track is obtained by successively locally averaging its previous scale. We get rid of coarsest levels and finest ones, as they respectively do not represent interesting features and are polluted by speckle noise (Fig. 1.4).

Fig. 1.4 Geo-projected reference and repeated sonar tracks before registration with patch A (red) and B (green)

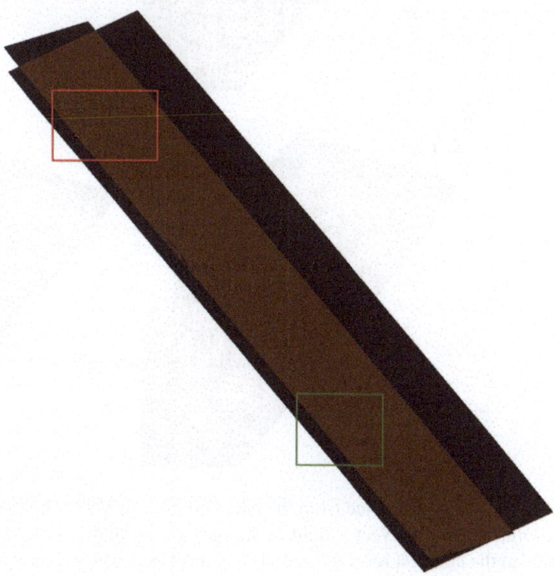

1.3.2 Results

Intermediate results of the registration procedure, using the parameters detailed in Table 1.1, are shown in Fig. 1.5 to illustrate the role of each step. Indeed, while the very first vector field computed at the coarsest scale contains several outliers (a), the navigation-based filtering is able to get rid of them as they do not match the navigation uncertainties provided by the INS. The remaining outliers (the two pink

Table 1.1 Block-matching semi-operational parameters used in our experiment

Parameters used in our experiment							
Resolution level	S_{ref} [m]	S_{rep} [m]	d	N	K	T_θ	$T_{\|\cdot\|}$
Coarsest	30	40	2	all vectors	5.991	45°	1.4
Intermediate	20	22					
Finest	15	16					

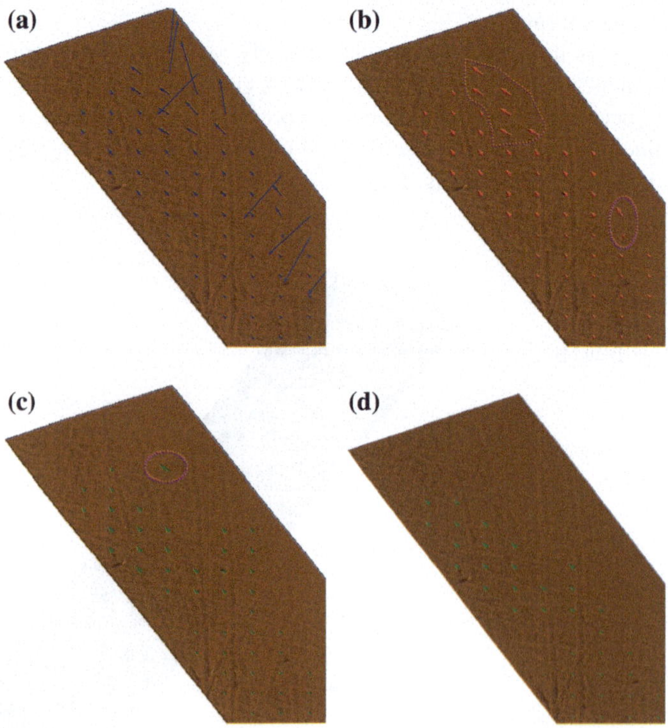

Fig. 1.5 Patch extracted from the repeated track. **a** Vector field after block-matching at the coarsest resolution level. **b** Vector field at the coarsest resolution after navigation-based filtering. **c** Vector field at the coarsest resolution after neighborhood-based filtering. **d** Filtered vector field at the finest resolution level

contours in (b)) are then filtered using the neighborhood-based approach described in Sect. 1.2.4. However, as it can be seen from (c), an outlier remains after all these filtering steps, but, as the correlation is probably not stable along the resolution levels, it disappears from the final vector field (d), and thus allows to obtain a more reliable warping of the repeated track.

We here provide both interpolated fields (X and Y axes) computed by means of Delaunay triangulation with cubic piecewise function and multilevel B-splines (Fig. 1.6) for the intermediary resolution level consisting in an image of 3072 $\times$ 3072 pixels. While the B-splines method provides a smoother field than the triangulation-based one (due to extrapolation outside the initial convex hull), our current implementation takes 28s to be computed whereas the triangulation-based one only needs 1.6s.

Nonetheless, as our sonar tracks are practically rendered using OpenGL, instead of computing the displacement for each pixel of the repeated track, we only compute such a displacement for the OpenGL vertices. Indeed, thus, when the vertices are shifted, the sonar track, which is stored as a texture in OpenGL, will be automatically warped through GPU interpolation, thus speeding-up the warping process.

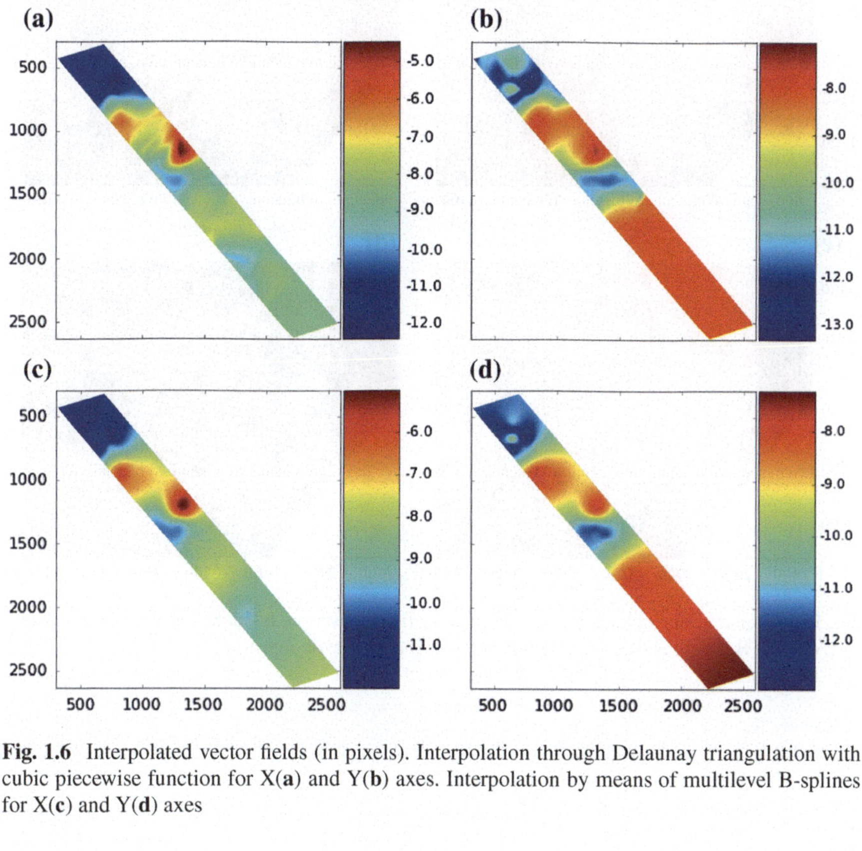

Fig. 1.6 Interpolated vector fields (in pixels). Interpolation through Delaunay triangulation with cubic piecewise function for X(**a**) and Y(**b**) axes. Interpolation by means of multilevel B-splines for X(**c**) and Y(**d**) axes

From Fig. 1.7 (a, b), we can see the effectiveness of our approach through the comparison of the previously studied tile A, before and after registration. The three objects lying on the seabed all superimpose between the reference track and the registered repeated one. In the second snippet B (c,d), while the trawling net mark and the MILCO object superimpose well after registration, it is not the case with the moored mine. Indeed, while its echo is correctly registered, shadows do not superimposed as such a mine is much more sensitive to point of view and grazing angle variation due to its geometry.

Regarding the whole computation time, we need about one minute to elastically register our sonar tracks (corresponding to an area of 5000 $\times$ 5000 pixels at the finest resolution), although it could be easily speeded up using modern software approaches namely multithreading or GPU computation.

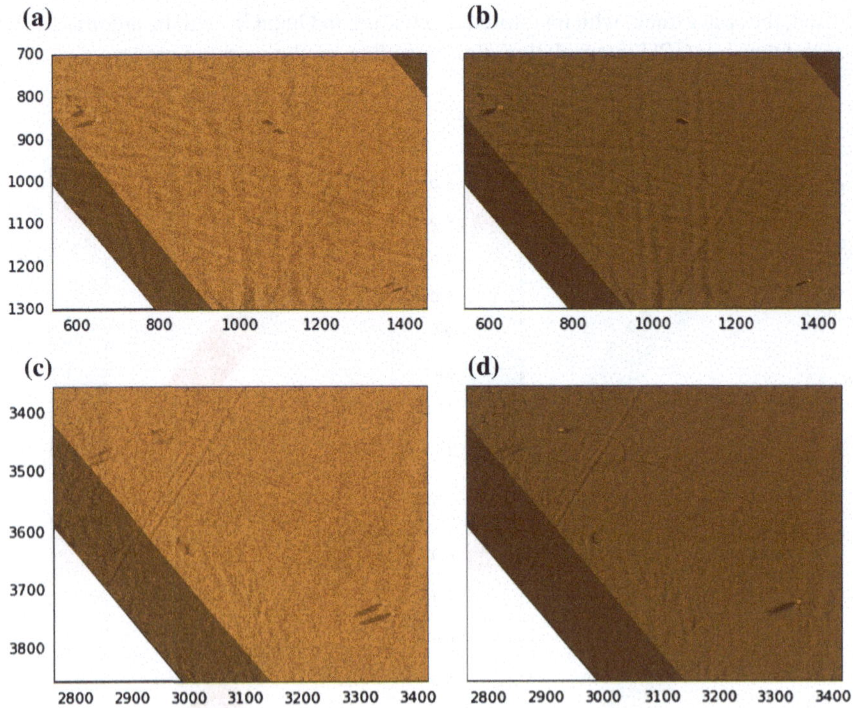

Fig. 1.7 Registration results. **a** Patch A before elastic registration. **b** Patch A after elastic registration. **c** Patch B before elastic registration. **d** Patch B after elastic registration

1.4 Conclusion

In this paper, we provide a method based on Fourier correlation to perform sonar tracks registration, along with a multi-resolution framework, allowing computational efficiency, and thus meeting our real-time constraint. Thanks to the data provided by the INS along with a neighborhood-based filtering, we get rid of outliers to finally obtain a vector field which can interpolated to then warp the repeated track. As our registration algorithm targets a change detection goal, where we try to accurately re-navigate the same area, the use of the Fourier correlation as the similarity metric should be sufficient to handle the expected geometric distortions such as grazing and/or view angle variation.

1.5 Further Work

Once both sonar tracks have been registered, it is conceivable to apply change detection algorithms to find out the areas where changes have occured.

References

1. B. Zerr, G. Mailfert, A. Bertholom, H. Ayreault, Sidescan sonar image processing for auv navigation, in *Oceans 2005-Europe*, (vol. 1), (June 2005), pp. 124–130
2. P. King, B. Anstey, A. Vardy, Comparison of feature detection techniques for auv navigation along a trained route. Oceans-San Diego **2013**, 1–8 (2013)
3. P. Vandrish, A. Vardy, D. Walker, O. Dobre, Side-scan sonar image registration for auv navigation, in *Underwater Technology (UT), 2011 IEEE Symposium on and 2011 Workshop on Scientific Use of Submarine Cables and Related Technologies (SSC)*, (April 2011), pp. 1–7
4. L. Bernicola, D. Gueriot, J.-M. Le Caillec, A hybrid registration approach combining slam and elastic matching for automatic side-scan sonar mosaic. Oceans-St John's **2014**, 1–5 (2014)
5. I. Leblond, Recalage Pà long terme d'images sonar par mise en correspondance de cartes de classification automatique des fonds, Ph.D. dissertation, (Université de Bretagne Occidentale, 2006)
6. C. Chailloux, J.-M. Le Caillec, D. Gueriot, B. Zerr, Intensity-based block matching algorithm for mosaicing sonar images. IEEE J. Ocean. Eng. **36**(4), 627–645 (2011)
7. J. Ferrand, N. Mandelert, Change detection for mcm survey mission, in *International Conference on Detection and Classification of Underwater Targets* (2012), pp. 193–206
8. I. Quidu, Incoherent change detection using amplitude sidescan sonar image, in *ECUA 2012* (2013). http://dx.doi.org/10.1117/12.2053067
9. I. Quidu, V. Myers, Ø. Midtgaard, R. Hansen, Subpixel image registration for coherent change detection between two high resolution sonar passes, in *iCoURS'12* (2012)
10. T. G-Michael, B. Marchand, J.D. Tucker, D.D. Sternlicht, T.M. Marston, M.R. Azimi-Sadjadi, Automated change detection for synthetic aperture sonar, vol. 9072, (2014), pp. 907 204–907 204–11. http://dx.doi.org/10.1117/12.2053067
11. D. Gueriot, E. Maillard, J.-P. Kernin, Sonar image registration through symbolic matching: a fuzzy local transform approach using genetic algorithms, in *OCEANS'96. MTS/IEEE. Prospects for the 21st Century. Conference Proceedings*, vol. 3, (Sep 1996), pp. 1324–1329

12. I. Barrodale, R. Kuwahara, R. Poeckert, D. Skea, Side-scan sonar image processing using thin plate splines and control point matching. Numer. Algorithms **5**(2), 85–98 (1993). http://dx.doi.org/10.1007/BF02212041
13. S. Daniel, F. Le Leannec, C. Roux, B. Soliman, E. Maillard, Side-scan sonar image matching. IEEE J. Ocean. Eng. **23**(3), 245–259 (1998)
14. J. Zhao, W. Tao, H. Zhang, K. Yang, Study on side scan sonar image matching based on the integration of SURF and similarity calculation of typical areas, in *OCEANS 2010 IEEE-Sydney* (May 2010), pp. 1–4
15. C. Rominger, A. Martin, A. Khenchaf, H. Laanaya, Sonar image registration based on conflict from the theory of belief functions, in *Information Fusion, 2009. FUSION'09. 12th International Conference on* (July 2009), pp. 1317–1324
16. M.T. Pham, D. Gueriot, Guided block matching for sonar image registration using unsupervised Kohonen neural networks, in *OCEANS 2013-San Diego: MTS/IEEE international conference* (2013), pp. 1–5
17. P.Y. Mignotte, M. Lianantonakis, Y. Petillot, Unsupervised registration of textured images: applications to side-scan **1**, 622–627 (2005)
18. C. Chailloux, B. Zerr, Non symbolic methods to register sonar images, in *Oceans 2005-Europe*, vol. 1 (June 2005), pp. 276–281
19. C. Chailloux, Recalage d'images sonar par appariement de régions : application à la génération d'une mosaïque, Ph.D. dissertation, UR1-Université de Rennes 1, ITI—Dépt. Image et Traitement Information (Institut Mines-Télécom-Télécom Bretagne-UEB), 2007, th. doct. : Traitement du signal et télécommunications, Université de Rennes 1, Institut Mines-Télécom-Télécom Bretagne-UEB (2007)
20. B. Bonnett, M. Hayes, Data-driven image registration for coherent change detection of synthetic aperture sonar imagery, in *Proceedings of the 29th International Conference on Image and Vision Computing New Zealand*, ser. IVCNZ'14, (New York, NY, USA: ACM, 2014), pp. 196–201. http://doi.acm.org/10.1145/2683405.2683455
21. P. Viola, W.M. Wells, III, Alignment by maximization of mutual information. Int. J. Comput. Vision **24**(2), 137–154 (1997). http://dx.doi.org/10.1023/A:1007958904918
22. J.P.W. Pluim, J.B.A. Maintz, M.A. Viergever, Mutual-information-based registration of medical images: a survey. IEEE Trans. Med. Imaging, 986–1004 (2003)
23. F. Nicolas, B. Zerr, A. Arnold Bos, Sonar image registration: survey and experiments in a rigid context (2016)
24. D. Padfield, Masked object registration in the fourier domain. IEEE Trans. Image Process. **21**(5), 2706–2718 (2012)
25. D. Ruprecht, H. Muller, Image warping with scattered data interpolation. IEEE Comput. Graph. Appl. 37–43 (1995)
26. D. Rueckert, L. Sonoda, I. Hayes, D. Hill, M. Leach, D. Hawkes, Nonrigid registration using free-form deformations: Application to breast mr images **18**(8), 712–21, 08 (1999)
27. Z. Xie, G.E. Farin, Image registration using hierarchical b-splines. IEEE Trans. Vis. Comput. Graph. **10**(1), 85–94 (2004)
28. S. Lee, G. Wolberg, S.Y. Shin, Scattered data interpolation with multilevel b-splines. IEEE Trans. Vis. Comput. Graph. **3**(3), 228–244 (1997). http://dx.doi.org/10.1109/2945.620490
29. A. Goshtasby, Piecewise cubic mapping functions for image registration. Pattern Recogn. **20**(5), 525–533 (1987) http://dx.doi.org/10.1016/0031-3203(87)90079-3

Chapter 2
Adaptive Sampling with a Fleet of Autonomous Sailing Boats Using Artificial Potential Fields

Hadi Saoud, Frédéric Plumet and Faiz Ben Amar

Abstract Autonomous sailing boats are an attractive solution to perform measurements at sea thanks to their low energy consumption. A fleet of such vehicles can also be used for adaptive sampling of an environmental field. In previous work, artificial potential fields have been used successfully in a local and real-time planner to carry out the autonomous navigation of one sailing boat. In this paper, we first extend the previous method to include marine current in the local planning and show how this extension can improve the safety and navigation performances in the case of strong marine current. Next we show how the same planning method can be used to control a formation of sailboats and perform adaptive sampling of some natural quantities and can be easily tuned to exhibit different behaviors such that way-point reaching, gradient or isoline following.

2.1 Introduction

Oceans observation at mesoscale (10–100 km) or sub-mesoscale (1–10 km) is an important issue for researcher in order to better understand how the marine environment evolves. However, at these scales, satellite remote sensing provides neither enough details on the sea surface nor provides information on the water column structure and these observation must be complemented by measures carried out by drifting profilers, research vessels, moored or drifting buoys for examples. Moreover, in marine environment, physical, chemical or biological fields have usually

H. Saoud · F. Plumet (✉) · F. Ben Amar
Sorbonne Universités, UMR 7222, ISIR, UPMC Univ Paris 06, 75005 Paris, France
e-mail: plumet@isir.upmc.fr

H. Saoud
e-mail: saoud@isir.upmc.fr

F. Ben Amar
e-mail: amar@isir.upmc.fr

F. Plumet
Université Versailles St-Quentin, UMR 7222, ISIR, 75005 Paris, France

L. Jaulin et al. (eds.), *Marine Robotics and Applications*, Ocean Engineering & Oceanography 10, https://doi.org/10.1007/978-3-319-70724-2_2

high spatio-temporal variability. It is so often desirable to perform adaptive sampling of these quantities at a scale compatible with the observed structure [2].

With the development of smaller and lower cost sensors, new solutions using autonomous robots (surface or underwater) begin to emerge in order to carry these sensors where measurements are more likely to be useful [8, 15, 18, 19, 27]. Using such low cost vehicles, adaptive sampling can thus be performed by following the gradient of some physical, chemical or biological parameter. To perform such gradient-based navigation in an efficient way (time constraints), a fleet of vehicles is usually required, even if it remains possible using a single vehicle (see [3] for example). Examples of gradient following navigation for a fleet of vehicles can be found in [1, 9, 14, 20].

Focusing on Autonomous Surface Vehicles (ASV), numerous robots have been developed for bathymetric data recording in shallow water [16], monitoring of various marine environmental data either alone [4, 17, 26] or as part of a sensor network [6, 7, 21, 24]. However, all of these aforementioned vehicles are conventional electrically powered systems and suffer from a lack of autonomy that restrains their usage to short-term missions. Autonomous sailing robots could then be an efficient solution for mesoscale oceans observation since they rely on renewable energy such as solar and wind energies [5, 25].

However, the control of autonomous sailboats is faced with two inherent difficulties: the uncontrollable and partially unpredictable nature of thrust forces (wind direction and wind speed) on one hand and the complex kinematics of a sailboat (aero and hydrodynamic properties of sails and hull) on the other hand. In previous work, we used artificial potential fields method [12, 13] in a local path planning algorithm that address these two problems simultaneously. This navigation method is able to react in real-time to changes in the environment and control the vehicle towards a given waypoint [22].

In this paper, we use the same framework of artificial potential field for gradient climbing of a natural field using a formation of autonomous sailboats. For the sake of clarity, the method used to control the motion of one single sailboat is first presented in Sect. 2.2. Since we focus on oceans parameters measurements, it is also necessary to consider the existence of ocean currents. Obviously, if strong current exists, the sailboat will drift and may not be able to reach it's final destination. Section 2.3 shows how to address these currents in the local path planning algorithm and how this extension can enhance the performance and safety of navigation in the case of strong marine currents. We made the assumption that these currents are known either through a marine chart or by local measurements made by the sailboat. Section 2.4 presents how adaptive sampling of a given natural field by a formation of autonomous sailboats easily fits in the general framework of potential field planning. The proposed method is based on the definition of a reference formation. The motion of the reference formation as well as the motion of the real vehicles are controlled by a local potential field planner i.e. each real vehicle is attracted by a potential field attached to the corresponding reference vehicle in the reference formation. The assumption is made that there exists a bidirectional communication link between the different vehicles to share the current and desired positions of the real vehicles as well as the position of the reference formation. Simulation results are presented and shows that this method can easily be tuned to exhibit different behavior such that way-point reaching for an entire formation, gradient or isoline following of a natural potential field.

2.2 Potential Field Planning of One Single Sailboat

Potential field (PF) method can be an efficient way to achieve local path planning for mobile robots. Since this method is a local one, its main drawback is that the moving vehicle may fall into a local minimum when the obstacle density is high. Luckily, open sea rarely have a high density of obstacles, making the potential field method a very convenient path planning algorithm for autonomous sailboats [10, 11, 23].

The fact that this method is local is also an advantage because the computational burden remains the same even if the obstacles or the goal are moving. This property was used to design a local planner for an autonomous sailboat, with the kinematic constraints of the sailboat transformed in obstacles moving with the vehicle. This property will also be used to design the local path planning algorithm for a formation of sailboat, with a moving goal (attractive potential) attached to each node of the reference formation (see Sect. 2.4).

In potential field based planning method, the vehicle has to follow the direction defined by the gradient of a scalar field. For autonomous sailboat path planning, the field is the summation (see Fig. 2.1) of several exo-potential fields and an endo-potential field:

- exo-potential fields are repulsive potentials centered at each obstacles position and an attractive field centered at the goal position,
- endo-potential field is a repulsive potential, moving with the sailboat and depending on the actual wind condition, that reflects the sailboat kinematics.

In practice, the gradient is evaluated numerically on a vehicle-centered ring at each sample time, using the actual value of the wind (speed and direction) measured by the sailboat and the position of a given way-point. Thus, at each time instant, the minimum of the local gradient defines the desired heading of the sailboat.

Notice that, since the desired heading is defined by the weighted sum of several potentials, the attractive potential must be defined as $U_g = K_g\, d_{goal}$ to guarantee that the sailboat will not tack too frequently or go into a no-go zone when the distance to the goal d_{goal} is important.

The endo-potential is built on the basis of the speed polar diagram of the sailboat (see Fig. 2.2). For a given wind speed, this polar curve shows the boat speed (in steady state) as a function of the angle γ between the heading and the true wind vector. Two high repulsive potentials are used to represent the upwind and downwind no go-zones.

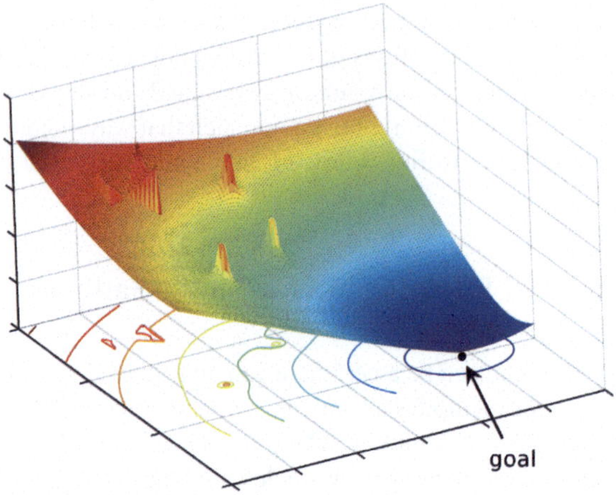

Fig. 2.1 Overall potential relative to three obstacles and a goal point (waypoint)

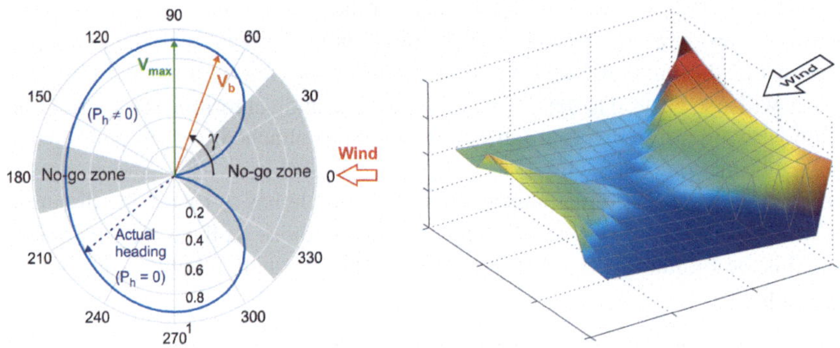

(a) Speed polar diagram of a sailboat (b) Close view of the endo-potential around the boat

Fig. 2.2 Endo-potential around the boat based on the speed polar diagram

Another potential P_w is added to reflect the speed performances of the sailboat. This potential can be defined as:

$$P_w = G_w \frac{V_{Max} - V_b}{V_{Max}} + P_h$$

with V_b the boat speed, V_{Max} the maximum boat speed and G_w a given weighting factor. In order to take the cost of gybing and tacking into account, an hysteresis potential $P_h \neq 0$ is added for the navigation area on the other side of the wind, relative to the actual heading.

2.3 Including Marine Currents

For application such that oceans parameters measurements with sailboats, it can be also mandatory to take ocean currents into account. Obviously, in such a case, the sailboat will drift and may not be able to reach it's final destination if the current is strong enough. We assume here that these ocean currents are known either through a marine chart or by local measurements made by the sailboat.

Since the local path planning algorithm is based on the speed polar diagram, we can easily include the marine currents by modifying the speed polar diagram.

Let $\overset{\to\mathcal{I}}{V}_b$ be the boat velocity vector in the inertial frame $\mathcal{I}$, $\overset{\to\mathcal{C}}{V}_b$ the boat velocity with respect to the current frame $\mathcal{C}$ and $\overset{\to\mathcal{I}}{V}_c$ the current velocity with respect to the inertial frame $\mathcal{I}$. Using the same notation for the wind vector, we can write:

$$\overset{\to\mathcal{I}}{V}_b = \overset{\to\mathcal{C}}{V}_b + \overset{\to\mathcal{I}}{V}_c \quad \text{and} \quad \overset{\to\mathcal{I}}{V}_w = \overset{\to\mathcal{C}}{V}_w + \overset{\to\mathcal{I}}{V}_c$$

Assuming that, for a given wind speed, the boat velocity vector $\vec{V}_b$ is given by a polar function $P()$ depending on the true wind vector $\vec{V}_w$ and angle γ between these two vectors (see Fig. 2.2a):

$$\vec{V}_b = P\left(\vec{V}_w, \gamma\right)$$

This relationship remains the same if both velocity vectors are expressed in the same frame. Since:

$$\overset{\to\mathcal{C}}{V}_b = P\left(\overset{\to\mathcal{C}}{V}_w, \gamma\right)$$

the boat velocity in the inertial frame is given by:

$$\overset{\to\mathcal{I}}{V}_b = \overset{\to\mathcal{C}}{V}_b + \overset{\to\mathcal{I}}{V}_c \;=\; P\left(\overset{\to\mathcal{I}}{V}_w - \overset{\to\mathcal{I}}{V}_c, \gamma\right) + \overset{\to\mathcal{I}}{V}_c \tag{2.1}$$

The resulting speed polar diagram (taking into account the effect of the marine current) is then the result of a translation (due to the $\vec{V}_c$ vector) and a slight deformation due to the non-linear effect of the $P()$ function. This is illustrated Fig. 2.3 for $\overset{\to\mathcal{I}}{V}_w = [-2, 0]^t\ (m/s)$ and $\overset{\to\mathcal{I}}{V}_c = [-0.25, 0.25]^t\ (m/s)$.

This modified speed polar diagram can then be used to build the adapted endo-potential around the sailboat. An example, showing the results of a local path planning, is depicted Fig. 2.4 using either the conventional speed polar polar diagram (see Fig. 2.4a) or the adapted version to take currents into account (Fig. 2.4b). For these simulations, the wind speed is equal to 4 m/s, the current speed is equal to 0.5 m/s and their directions are depicted Fig. 2.4.

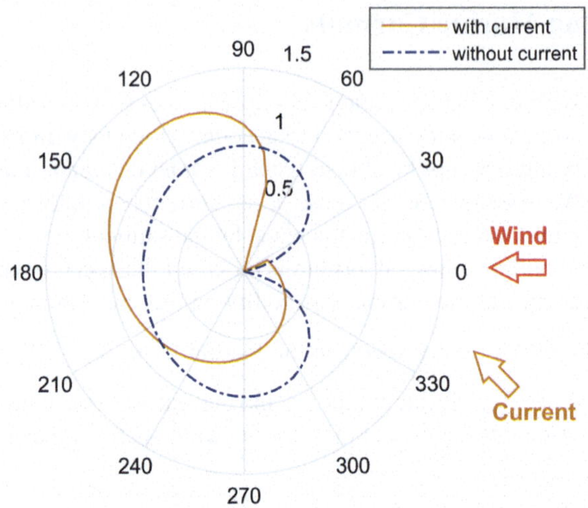

Fig. 2.3 Speed polar diagram with marine current ($\overset{\rightarrow\mathcal{I}}{V}_w = [-2, 0]^t$ and $\overset{\rightarrow\mathcal{I}}{V}_c = [-0.25, 0.25]^t$)

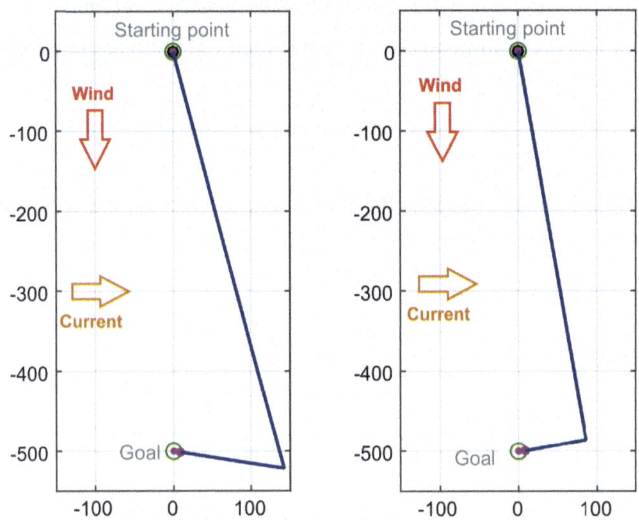

(a) Resulting path, without taking current into account

(b) Resulting path, taking current into account

Fig. 2.4 Potential field based planning with and without taking into account oceans current

One can see that, when taking currents into account in the path planning algorithm, the boat follows a more direct route toward the goal. Moreover, the time to reach the waypoint is equal to 9.1 mn when taking currents into account and equal to 10.8 mn when using the conventional method, which corresponds to an enhancement of more than 10% to the total traveling time.

2.4 Formation Control

As mentioned before, the general framework of potential field based planning can easily be adapted to solve the problem of gradient following of a natural field with a fleet of sailboats. Moreover, since potential field algorithm is a local path planning method, the computational burden remains the same if the goal is moving. This property will be used to control a formation of sailboats.

The basic idea is to define first a reference formation where each node of the reference formation becomes an attractive goal for one corresponding real sailboat. Then, the whole reference formation is steered toward a given waypoint or driven to follow the gradient of a natural potential field. In this way, the vehicles remains in formation but can adapt their motion to the local wind conditions.

We first define the reference formation as a rigid lattice, with each node of the lattice corresponding to a reference (virtual) vehicle. Let ${}^{(i)}S_{Ref}$ be the position of the (i) reference vehicle and R a local frame attached to the reference formation. Without loss of generality, we assume here a triangular formation with a local frame R attached to the first node of the formation (see Fig. 2.5a).

The motion of each real sailboat (i) is then driven by a force ${}^{(i)}\vec{F}$ coming from the overall potential ${}^{(i)}U$:

$$ {}^{(i)}\vec{F} = \overrightarrow{grad}\left({}^{(i)}U\right) = {}^{(i)}\vec{F}_g + {}^{(i)}\vec{F}_o + {}^{(i)}\vec{F}_w \tag{2.2} $$

with ${}^{(i)}\vec{F}_g$ the attractive force coming from the (i) reference vehicle, ${}^{(i)}\vec{F}_o$ the force due to external obstacles as well as other real vehicles and ${}^{(i)}\vec{F}_w$ the force coming from the endo-potential of the speed polar diagram, including marine currents if present. This way, each real sailboat is linked to a moving reference node in the reference formation (see Fig. 2.5b) but remains free to locally adapt its motion to varying wind and current conditions as well as avoiding collision with other vehicles of the formation (or external obstacles). These forces, as well as the positions of the

Fig. 2.5 Definition of the reference formation with the corresponding real vehicles

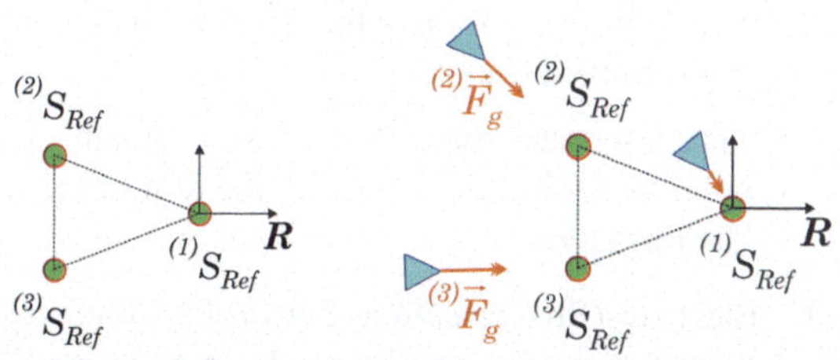

(a) Reference formation, with its attached frame R

(b) Reference formation with the "linked" real vehicles

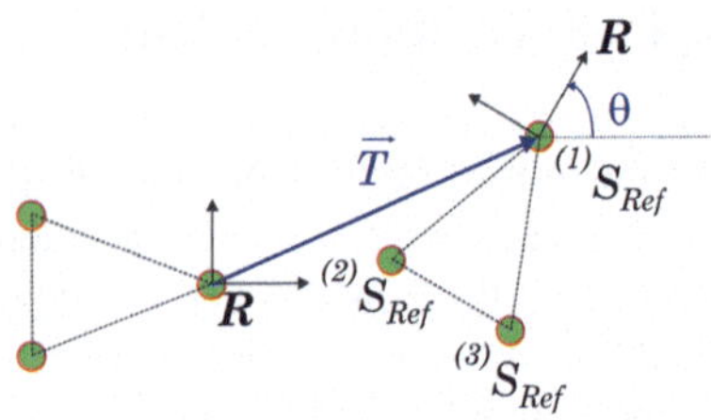

Fig. 2.6 Displacement of the reference formation

reference nodes are computed at each sampling time T_s of the local path planning algorithms implemented on each vehicle.

The last step is the displacement of the whole reference formation. This displacement is defined by a translation $\vec{T}$ and a rotation θ of frame R with respect to the previous position of the frame (see Fig. 2.6).

- The norm of the translation vector can be used to perform a cruise control of the whole formation and ensures that each vehicle remains in the vicinity of its corresponding reference node and can be computed, for example, as:

$$\left|\vec{T}\right| = Max\left(V_d T_s - K_v \overline{dist}, 0\right)$$

  with V_d the desired velocity of the reference formation, $\overline{dist}$ the mean distance between the sailboats and their corresponding reference positions and K_v a positive scalar.

The direction of this vector is chosen accordingly to the mission objective: reach a waypoint, follow the gradient or an isoline of a natural field (computed locally by the fleet of sailboats). This direction can be computed either directly (for gradient or isoline following tasks for example) or by using a potential field path planning algorithm to drive the whole reference formation toward a feasible direction (for waypoint reaching for example).

- The orientation θ can be set arbitrarily.
  For example:
  - if $\theta = 0$, the orientation of the reference formation remains constant with respect to a given fixed frame,
  - if θ is chosen such that $\overrightarrow{{}^{(2)}S_{Ref}\,{}^{(3)}S_{Ref}}$ remains perpendicular to $\vec{T}$, the formation is pointing toward the goal with the first reference vehicle becoming the leader of the formation.

In order to test this formation control algorithm under varying wind angle and wind speed, a simulation environment has been designed. This environment is defined by a constant (mean) value of wind speed and wind angle as well as weather disturbance areas with a Gaussian distribution around a maximum value (see Fig. 2.7 for an example).

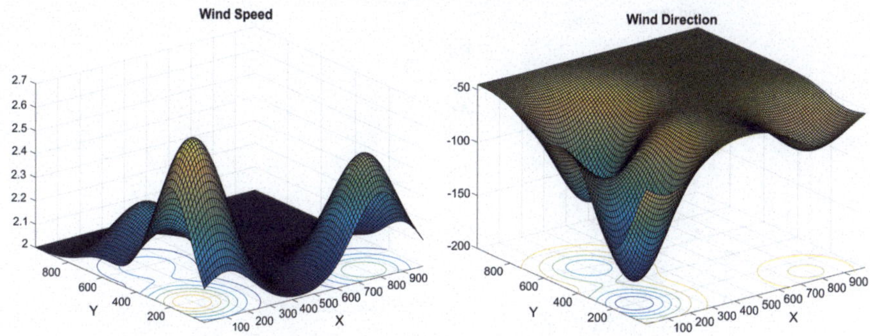

(a) Wind speed (b) Wind angle

Fig. 2.7 Wind field in the area

In the following simulations, the marine current was supposed to be null and the following simplified kinematic model of the sailboat was used:

$$\begin{cases} \dot{x} = u\ \cos(\psi) \\ \dot{y} = u\ \sin(\psi) \\ u = \left|\overrightarrow{V}_b\right| = \left|P(\overrightarrow{V}_w, \psi - \phi)\right| \end{cases}$$

with ψ the boat heading (defined by PF algorithm) and ϕ the wind angle.

Figure 2.8 shows the simulation result of a mission consisting in reaching a waypoint on $[300, 400]\ m$, starting from $[100, 100]\ m$. In this case, the fleet is constrained to stay in a leader-follower formation (for the reference formation, θ is chosen such that $\overrightarrow{^{(2)}S_{Ref}\ ^{(3)}S_{Ref}}$ remains perpendicular to $\overrightarrow{T}$) and the PF planning is applied to the leader of the reference formation (origin of the moving frame R). One can notice that the formation leader (as well as its "linked" real sailboat) is going directly to the goal while the second vehicle has to perform maneuvers (tacking) due to the local wind conditions in order to reach the waypoint.

The same control algorithm can be used for adaptive sampling of a natural parameter. In this case, each sailboat is supposed to be able to measure the given parameter and the hypothesis is also made that, given these measurements at each sample time, a representative gradient of the field can be estimated. The previous method can be readily used in this case except that a PF planning algorithm is no more used in order to compute the direction of motion of the reference formation (direction defined by the $\overrightarrow{T}$ vector). Instead, the gradient of the natural field is directly used to define the direction of $\overrightarrow{T}$.

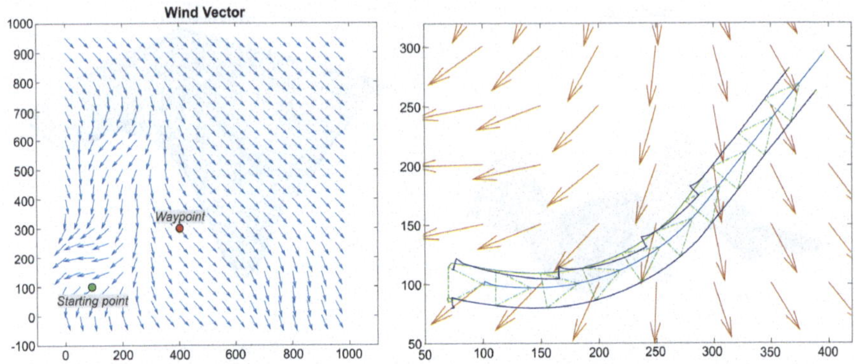

(a) Wind vector (b) Sailboats trajectories and local wind vector

Fig. 2.8 Waypoint reaching: sailboats and reference formation trajectories

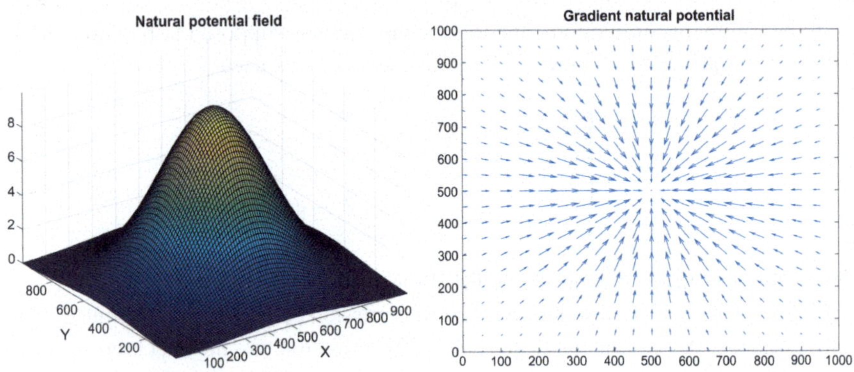

(a) Natural field (b) Gradient of the natural field

Fig. 2.9 Natural field and gradient of this field

Figure 2.9 shows an example of a natural field build on the basis of a Gaussian distribution around a central point located at the center of the map.

The mission depicted Fig. 2.10 consists of following the gradient of the natural field and then, when a given value of the field is reached, following the isoline contour of the field. One can notice that the sailboats are first moving upwind with several tacking maneuvers when following the gradient then follow a contour line at a constant value of the natural field (gradient of the natural field are depicted by arrows on Fig. 2.10b). In this case, we chose a constant orientation in a fixed frame for the reference formation ($\theta = 0$).

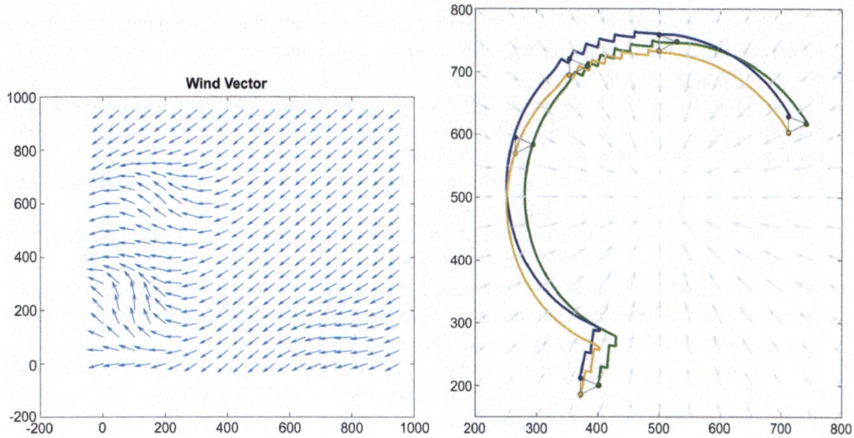

(a) Wind vector (b) Sailboats trajectories and gradient of the natural field

Fig. 2.10 Gradient then isoline following: wind vector and sailboats trajectories

2.5 Conclusion

This paper proposed a potential field planning method to control a formation of sailboats that can be used to perform adaptive sampling of a given natural parameter. The method is based on the definition of a rigid reference formation, with each sailboat linked to a node of this reference formation through an attractive potential. The effect of marine currents can be taken into account by modifying the speed polar diagram of the sailboat, which forms the core of the local planning algorithm. Several simulations show the effectiveness of the method for missions such that waypoint reaching, gradient or isoline following of a natural field under spatially varying wind conditions. This method can easily be implemented in real time on small embedded computer, provided that a bidirectional communication link exists between the sailboats for the coordinated motion control and exchange of local measurements of the natural parameter to build a local estimation of the natural field gradient.

References

1. T. Adamek, C.A. Kitts, I. Mas, Gradient-based cluster space navigation for autonomous surface vessels. IEEE/ASME Trans. Mech. **20**(2), 506–518 (2015)
2. A. Alvarez, B. Garau, A. Caiti, Combining networks of drifting profiling floats and gliders for adaptive sampling of the ocean, in *IEEE International Conference on Robotics and Automation* (2007), pp. 157–162

3. E. Burian, D. Yoerger, A. Bradley, H. Singh, Gradient search with autonomous underwater vehicles using scalar measurements, in *Symposium on Autonomous Underwater Vehicle Technology* (1996), pp. 86–98
4. M. Caccia, R. Bono, G. Bruzzone, E. Spirandelli, G. Veruggio, A.M. Stortini, G. Capodaglio, Sampling sea surfaces with SESAMO: an autonomous craft for the study of sea-air interactions. IEEE Robot. Autom. Mag. **12**(3), 95–105 (2005)
5. N.A. Cruz, J.C. Alves, Autonomous sailboats: an emerging technology for ocean sampling and surveillance, in *MTS/IEEE OCEANS* (2008)
6. J. Curcio, J. Leonard, A. Patrikalakis, SCOUT-a low cost autonomous surface platform for research in cooperative autonomy, in *MTS/IEEE OCEANS*, vol. 1 (2005), pp. 725–729
7. D.P. Eickstedt, M.R. Benjamin, H. Schmidt, J.J. Leonard, Adaptive control of heterogeneous marine sensor platforms in an autonomous sensor network, in *IEEE/RSJ International Conference on Intelligent Robots and Systems* (2006), pp. 5514–5521
8. C.C. Eriksen, T.J. Osse, R.D. Light, T. Wen, T.W. Lehman, P.L. Sabin, J.W. Ballard, A.M. Chiodi, Seaglider: a long-range autonomous underwater vehicle for oceanographic research. IEEE J. Ocean. Eng. **26**(4), 424–436 (2001)
9. E. Fiorelli, P. Bhatta, N.E. Leonard, I. Shulman. Adaptive sampling using feedback control of an autonomous underwater glider fleet, in *13th International Symposium on Unmanned Untethered Submersible Technology (UUST)* (2003)
10. L. Jaulin, F. Le Bars, A simple controller for line following of sailboats. Int. Robot. Sailing Conf. 117–129
11. L. Jaulin, F. Le Bars, An interval approach for stability analysis: application to sailboat robotics. IEEE Trans. Robot. **29**(1), 282–287 (2013)
12. O. Khatib, Real-time obstacle avoidance for manipulators and mobile robots. Int. J. Robot. Res. **5**(1), 90–98 (1986)
13. B. Krogh, C. Thorpe, Integrated path planning and dynamic steering control for autonomous vehicles. IEEE Int. Conf. Robot. Autom. **3**, 1664–1669 (1986)
14. F. Lekien, L. Mortier, P. Testor, Glider coordinated control and lagrangian coherent structures, in *2nd IFAC Workshop on Navigation, Guidance and Control of Underwater Vehicles* (2008), pp. 125–130
15. N.E. Leonard, D.A. Paley, F. Lekien, R. Sepulchre, D.M. Fratantoni, R.E. Davis, Collective motion, sensor networks, and ocean sampling. Proc. IEEE **95**(1), 48–74 (2007)
16. P. Mahacek, T. Berk, A. Casanova, C. Kitts, W. Kirkwood, G. Wheat, Development and initial testing of a SWATH boat for shallow-water bathymetry, in *MTS/IEEE OCEANS* (2008)
17. P. Mahacek, R. Kobashigawa, A. Schooley, C. Kitts. The WASP: an autonomous surface vessel for the university of alaska, in *MTS/IEEE OCEANS*, vol. 3 (2005), pp. 2282–2291
18. J.E. Manley, Unmanned surface vehicles, 15 years of development, in *MTS/IEEE, OCEANS* (2008), pp. 1–4
19. J.E. Manley, S. Willcox, The Wave Glider: a new concept for deploying ocean instrumentation. IEEE Instrum. Meas. Mag. **13**(6), 8–13 (2010)
20. P. Ogren, E. Fiorelli, N.E. Leonard, Cooperative control of mobile sensor networks:adaptive gradient climbing in a distributed environment. IEEE Trans. Autom. Control **49**(8), 1292–1302 (2004)
21. A. Pascoal, P. Oliveira, C. Silvestre, et al., Robotic ocean vehicles for marine science applications: the european ASIMOV project, in *MTS/IEEE OCEANS 2000*, vol. 1 (2000), pp. 409–415
22. C. Pêtrès, M.A. Romero-Ramirez, F. Plumet, A potential field approach for reactive navigation of autonomous sailboats. Robot. Auton. Syst. **60**(12), 1520–1527 (2012)
23. F. Plumet, C. Petres, M.-A. Romero-Ramirez, B. Gas, Sio-Hoi Ieng, Toward an autonomous sailing boat. IEEE J. Ocean. Eng. **40**(2), 397–407 (2015)
24. G.W. Podnar, J.M. Dolan, A. Elfes, S. Stancliff, E. Lin, J.C. Hosier, T.J. Ames, J. Moisan, T.A. Moisan, J. Higinbotham, E.A. Kulczycki, Operation of robotic science boats using the telesupervised adaptive ocean sensor fleet system, in *IEEE International Conference on Robotics and Automation* (2008), pp. 1061–1068

25. P.F. Rynne, K.D. von Ellenrieder, A wind and solar-powered autonomous surface vehicle for sea surface measurements, in *MTS/IEEE OCEANS* (2008)
26. E.T. Steimle, M.L. Hall, Unmanned surface vehicles as environmental monitoring and assessment tools, in *MTS/IEEE OCEANS* (2006)
27. L. Yongkuan, Auv's trends over the world in the future decade, in *Symposium on Autonomous Underwater Vehicle Technology* (1992), pp. 116–127

Chapter 3
Underwater Robots Equipped with Artificial Electric Sense for the Exploration of Unconventional Aquatic Niches

Stéphane Bazeille, Vincent Lebastard and Frédéric Boyer

Abstract This article presents different use of the electric sense perception in the context of underwater robot navigation. To illustrate the developed navigation behaviours we will introduce a recently launched european project named subCULTron and will show some simulation and experimentation results. The project subCULTron aims at achieving long-term collective robot exploration and monitoring of underwater environments. The demonstration will take place in the lagoon of Venice, a large shallow embayment composed of salt turbid water that represents a challenging environment for underwater robots as common sensors like vision or acoustic are difficult to handle. To overcome turbidity and confinement problems our robots will be equipped with artificial electric sensors that will be used as the main sensorial modality for navigation. Electric sense is a bio-inspired sense that has been developed by several species of fish living in turbid and confined underwater environment. In this paper, many different robotic behaviours based on the electric field perception will be presented, in particular we will address reactive navigation, object/robots detection, and object localization and estimation.

3.1 The subCULTron Project

3.1.1 Working Context

Water covers roughly 70% of the planet but is still mainly unknown due to its difficult exploration by humans (vast size, changing light conditions, turbidity). However, the underwater habitats has a high-impact for climate and ecological balance and requires the development of new technologies for robots. Within the subCULTron project we aim at developing new robotic technologies for the collective exploration and the monitoring of underwater environments. SubCULtron (submarine cultures perform

S. Bazeille (✉) · V. Lebastard · F. Boyer
Laboratoire des Sciences du Numérique de Nantes (LS2N),
Institut Mines Télecom Atlantique, Nantes, France
e-mail: stephane.bazeille@gmail.com

L. Jaulin et al. (eds.), *Marine Robotics and Applications*, Ocean Engineering & Oceanography 10, https://doi.org/10.1007/978-3-319-70724-2_3

long-term robotic exploration of unconventional environmental niches) is a project supported by European Union Horizon 2020 research and innovation program.[1] Its application will be focused on collecting large and long term environmental data (for instance: pressure, temperature, pH, salinity, conductivity, turbidity, chemical composition of water, water level, flow rate and pictures about the marine fauna and flora). These information will be in turn studied by biologists to understand the changes inherent to human activities in the region of Venice. The project is coordinated by the Artificial Life Lab of the University of Graz (Austria) and includes the Unit of Social Ecology of Université Libre de Bruxelles (Belgium), Cybertronica Research Center of Advanced Robotics and Environmental Science (Germany). The Laboratoire des Sciences du Numérique de Nantes (France), the Biorobotics Institute at Scuola Superiore Sant Anna (Italy), the Faculty of Electrical Engineering and Computing at University of Zagreb (Croatia), the Consortium for coordination of research activities concerning the Venice lagoon system (Italy).

3.1.2 Swarm Exploration and Monitoring

The novelty of subCULTron concentrates on the data collection that will be performed by a large robot swarm constituted of more than 130 entities (a swarm is a group or aggregation of free-swimming organisms). This artificial robot organization will be composed of 3 classes of cooperating robots (see Fig. 3.1): 5 floating platforms called artificial lily pads (aPad), 100 artificial mussels (aMussel) and 30 artificial fish (aFish) that will collaborate together to perform their mission.

aPad: The aPads are floating robots (see Figs. 3.2, 3.3b). This class of autonomous robots globally localized by GPS and equipped with solar panels, will be used as geolocalization satellites and recharge/gathering station for the underwater robots. They will communicate with the underwater robots though acoustic sensing. aPads will also maneuver actively and they have the longest runtime.

aMussel: The aMussels is a class of underwater robots that sits on the ground (see Fig. 3.3a and b). These robots have an actuation limited to a buoyancy device. They will dive, and autonomously surface for recharge. They will be used under a low-energy consumption regime and they will harvest energy from the bacteria. These robots will self-distribute in the habitat, and then be used as landmark, and data storage station for aFish.

aFish: The aFish are the most active robots (see Fig. 3.4). aFish will browses the habitat and organize to explore an area and collect data. As the most active robot, they have to autonomously manage their energy and to recharge their battery on aPads when they need. They will communicate with aMussels and aPads to localize

---

[1] subCULTron (2015–2019), EU-H2020, FET PROACT-2-2014 under grant agreement no 640967. http://www.subcultron.eu.

Fig. 3.1 Illustration of an experiment using the 3 different robots developed within the subCULTron project

Fig. 3.2 An aPad in the Arsenale in Venice

Fig. 3.3 Left: An aMussel, Right: An aMussel recharging on an aPad

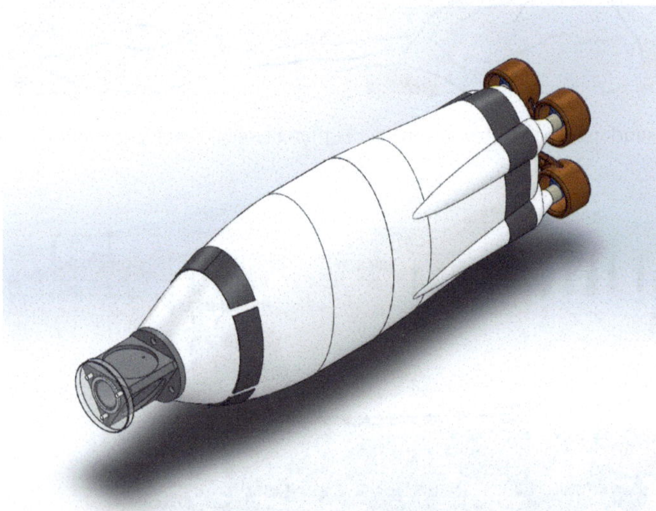

Fig. 3.4 First design of the aFish (artificial fish)

and will exploit information collected by other aFish in order to actively explore the environment.

All experiments will take place in Italy in the Venice Lagoon (a shallow water area of 500 km^2). This place is a particularly challenging environment for common underwater robots as it is a large heterogeneous area constituted of: a network of turbulent canals, open waters, mud flats, tidal shallows or salt marshes. It presents many difficulties that have not been addressed: a ubiquitous turbidity, a fluctuating salinity, potential strong water currents and a large variability of biological activities. As common sensing cannot be used under these conditions, its exploration pushed us towards the use of a new perception sensor: the electric sense.

3.1.3 *Robots Sensors Equipment*

In the following we will concentrate our discussion on the underwater robots: the aFish and the aMussel. The aPad as a floating platform will not be equipped with electric sense. The underwater robots will dispose of 2 different set of sensor. A set of sensors to collect data about the environment (turbidity, pH, temperature …) and navigation sensors. Among these navigation sensors, robots will dispose of common underwater sensors: an imu, an acoustic pinger and receiver, a camera, and a modulated light perception system. Besides this sensors underwater robots will include an artificial electric sensor that will be used as their main perception and communication sensor. Few reasons can explain this choice of electric sensing. First the omnipresence of suspended matters makes the water turbid or muddy, and the shallow waters rich in vegetation makes the environment cluttered and confined. These two specificities prevent robots from using common long and medium range underwater sensors such as sonar and vision. Vision, because the visibility is dependent on the available light energy which dramatically decreases in highly turbid turbulent and polluted waters. Sonar, because the reverberation of multiple echoes from the obstacles and the diffraction by suspended particles jam the sonar signals and makes their interpretation difficult. The electric sense giving an omnidirectional perception of the environment robust to the lack of light and the turbidity makes it perfectly suitable for our applications.

3.2 The Electric Sense in Nature

Electric sense has been observed almost exclusively in aquatic or amphibious animals, the known exceptions being echidnas, cockroaches and bees [8]. Underwater, several species of fish have the capacity to sense changes in electric fields in their vicinity [6]. Among fishes, we can distinguish two typical modes of electroreception: some fish passively sense changes in the nearby electric fields (passive electric sense) [10], some generate their own weak electric fields and sense the distortions of these fields with their skins (active electric sense) [7, 15].

3.2.1 *Passive Mode*

Passive electroreception is the most common modality: the fish senses the weak electric field generated by other animals (prey or conspecifics) and uses it to locate them. These electric fields are typically generated by nearly any species due to the activity of their nerves and muscles. Passive electric sense is used by several species of sharks and rays that have evolved specific electro-receptors named Lorenzini ampullae [10]. It has to be noted that these receptors are so sensitive that these fish also use them to orientate along the telluric fields or to sense the weak gradients of salinity. When hunting, the range of perception of passive electroreception is generally around few meters.

3.2.2 Active Mode

In active electric sense, the fish called weakly electric fish, can sense their nearby environment by detecting the distortions of a self generated electric field. This electric field is generated thanks to an electric organ that is located at the base of its tail. It generates a dipolar shaped electric field around the fish which is the distorted by the nearby objects as it shown on Fig. 3.6. Then the distortions is measured using a dense array of electro-receptors distributed over the fish skin. This phenomena called "electrolocation" can be compared to the echolocation used by dolphins, where the carrier is no longer an acoustic wave but an electric field. Because electric emitters are dipoles instead of sources, electric sense has a much shorter range (around 1 body length) but it has the advantage to be omnidirectional. They can also use electric sense for passive sensing to hunt their preys or to escape from their predators. Most of electric fish are principally nocturnal and live in confined turbid waters of the equatorial forests [14], that is to say waters that are rich in suspended particles with many obstacles such as the roots of the trees. Beyond perception, active electric fish can also communicate by modulating their electrical activity and they use this further ability for courtship behaviors or to mark out their territory. As an example, Fig. 3.5 shows the elephant-nose fish, the most clever electric fish. Recent behavioral experiments have shown that through active electric sense, this fish can localize preys, predators, conspecifics and also inert objects when they are electrically contrasted with respect to the water. Moreover, researchers showed that the fish can also identify

Fig. 3.5 A weakly electric fish: Gnathonemus petersii

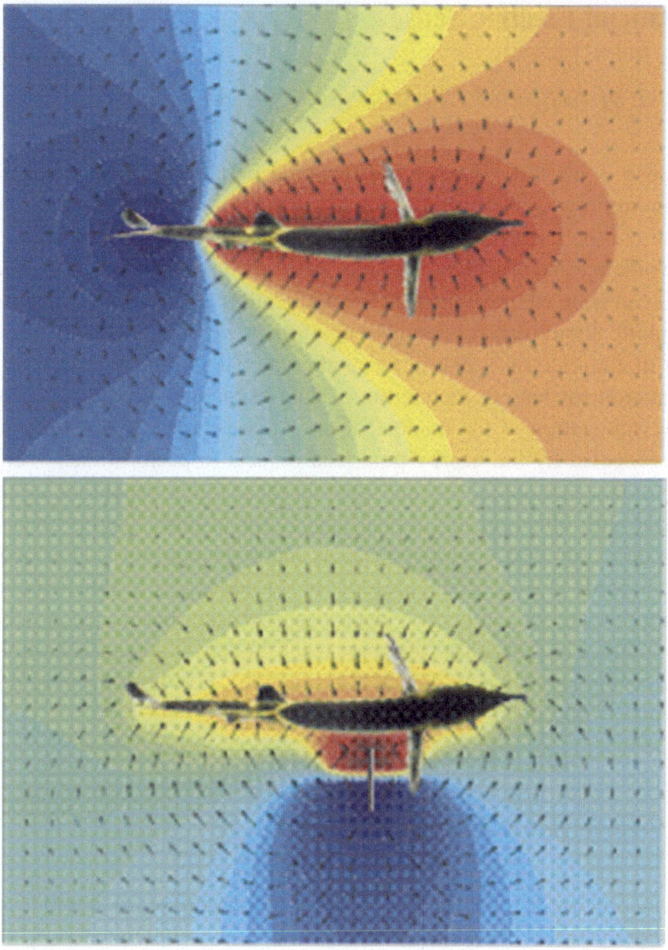

Fig. 3.6 Up. Basal electric field produced by the weakly electric fish. Down) Electric field re-emitted by the conductive polarized cylinder situation on the right side of the fish

the electric nature (conductive, insulating), and discriminate the sizes and shapes of objects [18]. Its polarization is performed in short pulses. Other fish of the family of *Gymnotide* use the same principle of reception but with a sinusoidal (alternative) electric field. These fish are named wave fish and their emission mode has inspired the artificial electric sensors [16].

3.3 Artificial Electric Sense

3.3.1 Use of Artificial Electric Sense for AUV

Electric sense is a short range omnidirectional perceptual ability well suited for confined environments and turbid or muddy waters. It is cheap, easy to integrate on a robot and can be used in different context such as reactive navigation, docking, obstacle avoidance and object recognition. Both active and passive modalities (active or passive) can be used depending on the application as the weakly electric fish do. When speaking about reactive navigation, a first example that clearly illustrates the advantage of the electric sense is the underwater docking of a passive robot on an active station for battery recharging or data exchange. Another possible use of the reactive navigation in the context of an active robot is to react to the electric field lines reflected by the nearby polarized objects to perform basic behaviors as obstacle avoidance or object seeking. Beyond these reflex behaviors, electric sense can be fused with other sensing modalities, as for instance inertial measurements to perform more complex cognitive tasks as shape recognition [11] or mapping the environment while self-localizing, a topic never addressed in the artificial electric sense community. Disadvantages of electric sense is the short range and the electric measures potentially difficult to interpret when the environment is complex (more than one emitting agent, many objects for example). For these reasons, it has to be complemented with other sensors.

3.3.2 Electric Sense in subCULTron

As discussed previously, electric sense being a short range sense, it will be supported with hydro-acoustic for long range localization, and complemented by modulated light at short range. Redundancy at short distance sensing has been chosen make the system more robust. Modulated light is an active sense as well. It is based on the emission of light that is reflected back by the obstacles and detected by some photoreceptors. Though being not much affected by the turbidity of the water, it can be severely disturbed by external light sources such as the sun when it is used close to the surface. In clear water, it has a little longer range than electric sense, typically about 1 meter but each photoreceptor has a small cone of perception then many devices are needed to obtain an omnidirectional perception without blind spots. Finally, 2 additional sensors will equip our underwater robots: an Inertial Measurement Unit (IMU) that gives the robot linear velocities and angular rates, and a pressure sensor that gives the depth. These proprioceptive information will be used for robot balancing (pitch, roll), control of the navigation (heading, velocity, depth). About electric sense, the two modalities will be used depending on the behaviours we need. Active mode will be extensively used to perceive the environment and navigate autonomously while avoiding obstacles. Passive mode will be used to navigate towards active robots (reach a base or follow a conspecific). In the following, we will consider passive electric sense as a particular case of active electric sense in which the electric field is no longer generated by the same sensor but by an external active dipole that can be another robot or a docking station for example. Passive electric sense will also be used as a way of communication between robots. However, the word communication will not refer to an exchange of an explicit information through coded messages, but will rather inform about the presence and position of others robots.

3.3.3 Basic Principle

Two artificial electric sensing technologies inspired by electric fish exist today $U-U$ [2] or $U-I$ [3, 17]. The first letter designates the first electric input controlling the electric emission (here a voltage U), the second, the measurement variable (I denotes a current). In both cases, the sensor is an insulating axisymmetric (plastic) shell on which a set of conductive electrodes are arrayed. The electric field is generated by setting a voltage on at least two electrodes in contact with the water. Though, both techniques ($U-U$ and $U-I$) share this common emission principle, in the $U-U$ mode, the other electrodes are paired floating potential electrodes between which the voltage is measured. While in the $U-I$ mode all of the electrodes except the emitter are grounded, and the currents that flow across each of them are measured. It is worth noting here, that till today artificial electric sense still remains restricted to tap waters and electric sensors designs were built for planar navigation. The sensor [17] will be used for some experiments presented in Sect. 3.5 is presented on Fig. 3.7. It is composed of an insulating shell with a set of 4 macro-electrodes denoted $\varepsilon_i, i = 0, 1, 2, 3$. The shell and the array of electrodes obey to a bilateral (left-right) symmetry. The back macro-electrode, ε_0 stands for the emitter, while the others $\varepsilon_1, \varepsilon_2, \varepsilon_3$ are the receivers. The emitter ε_0 is set under a controlled voltage U with respect to all the receivers which are grounded. This voltage U is imposed through a sine wave generator. Note here that a continuous voltage would generate an undesirable electrolysis. The surrounding water being a conductive medium, such a polarization of the sensor generates an electric field in the sensor surrounding (see Fig. 3.6). When there is no object within the robot's range, this electric field is named "the basal field".

All macro-electrodes are divided into 2 electrodes as shown on Fig. 3.7. The basal currents measurements are gathered in the vector $I^0 = [I_1, I_2, I_3, I_4, I_5, I_6, I_7, I_8]$. These currents are entirely modeled by the following the vector-equation:

$$I^{(0)} = C^{(0)}\overline{U}, \tag{3.1}$$

where $C^{(0)}$ is a 8×8 vector modeling the basal conductivity (indexed (0)) of the current paths between the emitter and the receivers, and $\overline{U}$ a vector of size 8×1 in which values are set to zero for the receiving electrodes and to U for the emitting ones. The currents measurement vector include 8 measurements, but it has to be noted that only 6 are useful because on the real robot due to an hardware limitation

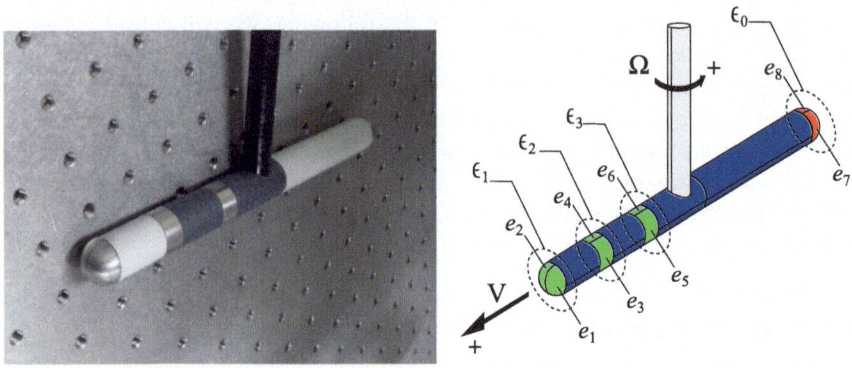

Fig. 3.7 Left: Artificial electric sensor [17], Right: Design of the sensor with our notations

the current cannot be measured on the emitting electrodes. The currents vector only depends on the sensor's geometry and the medium's conductivity γ_0 through the relation:

$$C^{(0)} = \gamma_0 S^{(0)}, \tag{3.2}$$

where $S^{(0)}$ is a matrix modeling the influence of the shape of the sensor on the conductance between the emitter and the receivers. For example, designing a sensor shape more and more complex, makes $S^{(0)}$ and so $C^{(0)}$ decrease since it obliges the electric current lines to be more and more curved. Another degree of freedom encoded into $S^{(0)}$ consists of the geometry of electrodes whose decreasing the size, makes $C^{(0)}$ decrease. Note also that $C^{(0)}$ can be obtained once for all either by numerical computation or in situ, through a preliminary calibration phase. Then, when an object (obstacle, other robot) appears in the sensor's surrounding, the vector of measured currents I becomes:

$$I = I^{(0)} + \delta I = (C^{(0)} + \delta C)U = CU, \tag{3.3}$$

where δI represents the perturbative component of the measured currents which images the presence of the object, and δC is its contribution to the external conductivity between the emitter and the receivers noted C. This perturbative conductivity δC depends on the geometry of the object, its position, as well as the dimension-less number $\lambda = \gamma/\gamma_o$, named contrast coefficient. In this respect note that when $\lambda = 1$, $\delta C = 0$, and the object is electrically transparent.

3.4 Electric Sense for the subCULTron

3.4.1 Electric Sensing in Salt Water

One of the main difficulty of the project is that we have to work in sea waters i.e. salt water, that is too say a water with a conductivity 100 higher than tap water. The conductivity defined as a measure of the capability of a medium to pass electric flow, depends on many parameters in particular the chemical compounds of water. Salinity has the strongest influence, the temperature has also a non negligible influence. As an example, tap water has a conductivity that ranges in $[0.005 - 0.05]$ S/m and salt water ranges in $[1 - 10]$ S/m. The conductivity in the lagoon of Venice is highly varying temporally and spatially, it has been estimated in [2–7] based on data collected in the lagoon in 2015. For our robots we preferred to use the U-I mode as it has a greater range equivalent to the fish range and it is more adapted to a use in sea water. Through this implementation we consider that the electromagnetic waves in the water are in the range of electric field frequencies ($\frac{w}{2\pi} < 50kHz$), and we restrict electric sense to a measure of the amplitude of electric current, the phase is neglected. As the conductivity is large and varying, the hardware presented in [17] has been augmented with a micro-controller to maintain a suitable emission with respect to the conductivity changes. When the conductivity increases the sensor saturates so the voltage U of the emitter has to be decreased, on the contrary when the conductivity decreases the amplitude has to be increased in order to keep the signal noise ratio and the perception range.

3.4.2 The Importance of the Sensor Morphology

For our 2 underwater robot, the morphology of the sensor has been also optimized in order to maximize the sensor range and maximize the information obtained from the current measures and ease the robot control. To maximize the sensor range we increased at the maximum the distance between the receivers and the emitters. On both robot the emitter and receivers are on opposite sides of the robots. Regarding the receiving electrodes we respected also a symmetric positioning constraint: aMussels are axisymmetric and aFish will have a left-right and up/down symmetry. Finally again to enhance the perception, the number of electrodes was chosen depending on the behaviours of each robot. For the aMussel which are vertically put on the seabed, we chosen to place the ring emitter at the bottom and 4 quarter of circle electrodes on the top as receivers to obtain directionality component from the current measures (see Fig. 3.8). For the aFish, we planned to have on a ring shape emitter on the tail and four electrodes on the nose (see Fig. 3.4). These receivers will be positioned in order to allow the detection of obstacles around the robot in 3D.

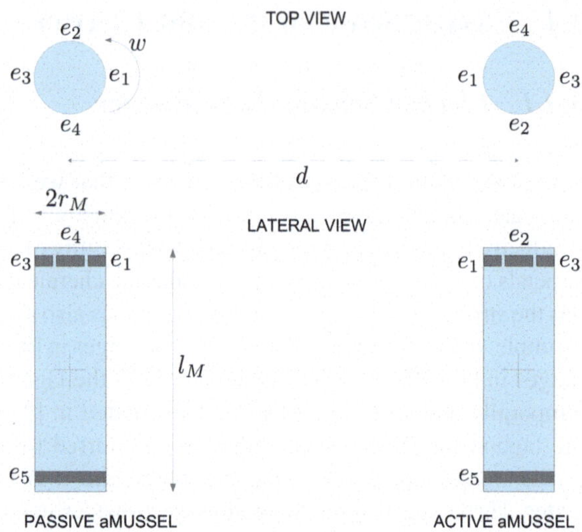

Fig. 3.8 aMussel experiment set up

3.5 Experiments Results Based on Electric Sense

For all the experiments described in this section we consider the aMussel is a cylinder with a length $l_M = 50$ cm and a diameter $r_M = 12$ cm (see Fig. 3.8). Regarding the aFish, that has still not be produced we used the slender probe [3] presented on Fig. 3.7. It has a length $l_F = 20$ cm, a diameter of $r_F = 2$ cm. We will show 4 different use of electric sense illustrated using the subCULTron underwater robots. Among the experiments shown, the aMussels experiment have been done in salt water, all other experiments have been only performed in tap water.

3.5.1 Localizing an Active Robot from a Passive Robot

In this first experiment, we consider a passive robot and an active robot in its range of perception in salt water. Our goal is to estimate the direction and the distance of the active robot (emitting an electric field) from the passive one. We only dispose of the 4 measured currents on the passive robot that linearly depends on both the conductivity γ_0, and the imposed voltage U. These two quantities are supposed unknown. To infer the distance and directionality, we computed the 2 following currents ratio from the 4 raw currents I_i (i = 1, 2, 3, 4) measured on the 4 electrodes e_1, e_2, e_3, e_4 (see Fig. 3.8):

$$R_{1,3} = \frac{I_1}{I_3} - 1, \qquad R_{2,4} = \frac{I_2}{I_4} - 1 \tag{3.4}$$

It has to be noted that computing the currents ratio makes our measurement independent from the conductivity. From these ratio we can estimate the directionality φ of the active robot defined by:

$$\varphi = ATAN2(R_{2,4}, R_{1,3}) \tag{3.5}$$

where φ is an angle that is defined a fixed reference frame attached to the passive robot. The direction estimated, we obtain the distance d using the following function:

$$d = 0.2\sqrt{(\ln R_{2,4})^2 + (\ln R_{1,3})^2} \tag{3.6}$$

This function is a parametrical model of the currents that has been computed in a preliminary phase from different measurements of the 2 currents ratio at different distances. As the currents ratio does not depend on the conductivity but only on the morphology of the aMussels and the distance between the 2 robots, the calibration can be done once for all in any water medium. To illustrate this localization of an active robot from a passive one in salt water we present on Fig. 3.9 the raw currents measured on the passive robot while rotating it in place (changing w on Fig. 3.8). It can be see that as w changes the maximum current value changes from electrodes 1–4. It is maximum when an electrode of the passive robot is facing the active robot. On Fig. 3.10, we show the currents measured on the passive robot as we pull it away from the active one but keeping the orientation fixed. The currents are in this case decreasing as the distance increases. It can be seen that the perception range of the sensor is about 0.8 m. To summarize, by using such a method we can estimate the robot distance and direction of an active robot from a passive one just by measuring 4 currents and without the knowledge on the conductivity or the voltage U.

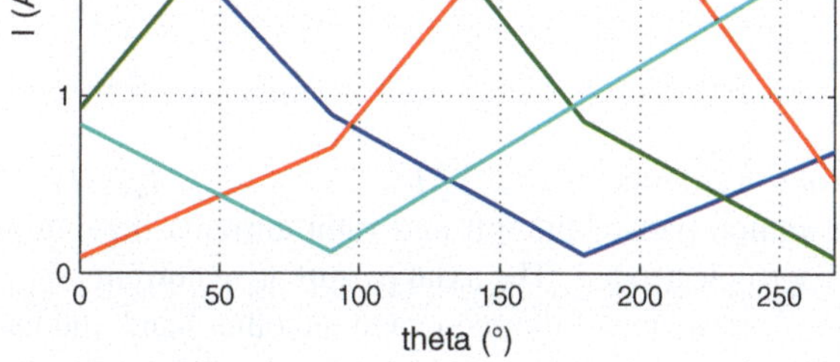

Fig. 3.9 Raw currents measured on the 4 electrodes of the passive aMussel with 4 different orientations (0°, 90°, 180°, 270°) at the same position

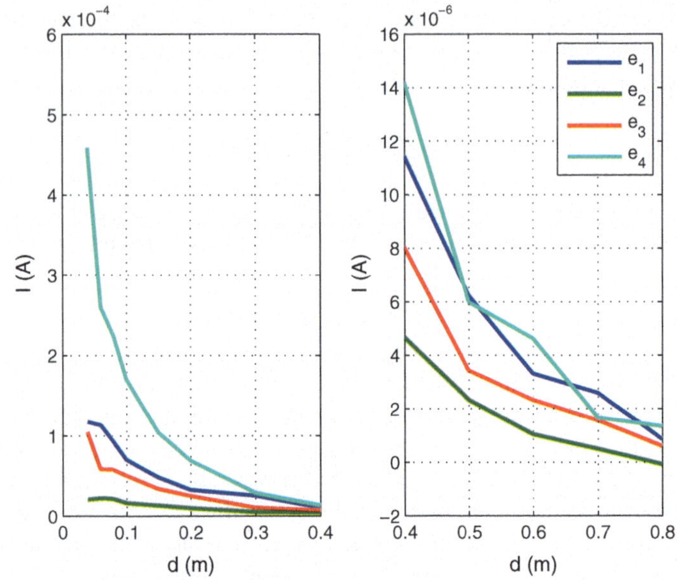

Fig. 3.10 Raw currents measured on the 4 electrodes of the passive aMussel while increasing the distance between the 2 aMussels

3.5.2 *Reactive Navigation Using Active Electric Sense (Memoryless)*

In this simulated experiment we suppose that we use the robot presented on Fig. 3.7. The macro-electrode ε_0 is set to a potential U supposed known. On each of these receiving electrodes the current is measured and we obtained the following measures: $I_1, I_2, I_3, I_4, I_5, I_6$. These values can be re-parameterized by the following 6 scalars:

$$\begin{aligned} I_{lat,1} &= \frac{(I_1-I_2)}{2} \\ I_{lat,2} &= \frac{(I_3-I_4)}{2} \\ I_{lat,3} &= \frac{(I_5-I_6)}{2} \\ I_{ax,1} &= \frac{(I_1+I_2)}{2} \\ I_{ax,2} &= \frac{(I_3+I_4)}{2} \\ I_{ax,3} &= \frac{(I_5+I_6)}{2} \end{aligned} \qquad (3.7)$$

And, we define $I_{lat} = \sum I_{lat,i}$, $I_{ax} = \sum I_{ax,i}$ $I_{ax,i}$ (axial current) represents the common part of the left and right currents flowing across the 2 electrodes of the macro-electrode i. The axial perturbative currents $I_{ax,i}$ is due to the variations of the total resistance of the scene. On the other hand, the lateral perturbative current $I_{lat,i}$ is proportional to the incident field. $I_{lat,i}$ (lateral current) represents the differential part of this left and right currents. This previous properties on $I_{ax,i}$ and $I_{lat,i}$ allows

Table 3.1 Sign of the perturbative axial and lateral currents δI_{ax} and I_{lat} with respect to the electric properties of the object and its side in the sensor frame. Currents were measured by the head electrodes when the object is in the frontal part of the sensor [5]

δI_{ax}	>0	for a conducting object
δI_{ax}	<0	for an insulating object
I_{lat}	>0	for a conducting object on the left of the sensor or for an insulating object on the right of the sensor
I_{lat}	<0	for a conducting object on the right of the sensor or for an insulating object on the left of the sensor
I_{lat}	$=0$	for any contrasted object facing the sensor

easily to determine some information of the environment. The axial current $I_{ax,i}$ is used to determine if the object is conductive or insulating. While the $I_{lat,i}$ currents is used to determine, with the knowledge on $I_{ax,i}$, if the object is on the left or on the right hand side of the sensor. When $I_{lat} = 0$, the sensor axis is necessarily aligned along the incident field (see Table 3.1).

Based on this 6 scalars defined in Eq. 3.7, by exploiting the morphology of the sensors (slender shape, bi-lateral symmetry), we can address the problem of navigation. A set of reactive control laws has been proposed [12], the principle is based on the alignment of the body of the robot-sensor on the electric lines emitted by the polarized object. Remarkably, these strategies are used by fish that hunt by following the electric lines emitted by their prey [9]. If we assume that the robot is controlled using 2 parameters: its linear velocity V and its angular velocity Ω. We can apply the following control law on V, Ω (see Fig. 3.7):

$$\begin{aligned} V &= C, \\ \Omega &= K\ I_{lat}, \end{aligned} \tag{3.8}$$

with C a constant positive value such that the sensor goes forward headlong and K a constant gain. By exploiting this control law, the robot just go straight at constant speed when there is not object in its surroundings but it will converge to an object or avoid an object depending on its electric properties (conducting and insulating) when such an object appears in its range of perception [5]. In more details the robot follows the electric line reflected by the polarized object as shown on Fig. 3.6 down. Depending on the object electric properties the measured values $\delta I_{ax} = I_{ax} - I(0)$ and I_{lat} varies following Table 3.1.

It is worth noting that δI_{ax} and I_{lat} have a complementary role since the former can be used to determine if the object is conductive or insulating, while the latter allows us to determine if it is on the left or on the right hand side of the sensor. With our convention on the sign of the currents and angular velocities, taking $K > 0$ in (Eq. 3.8) ensures that when a conductive object is on the right (respectively on the left), the sensor turns to the right (respectively to the left). On the other hand, if the object is insulating and on the right (respectively the left), the control law makes it react as if there was a symmetric conductive object on the left (respectively the right). Thus, the controller repulsed the sensor from an insulating object. To invert the behavior, i.e. makes the sensor attracted by insulating objects and repulsed by conductive ones, the sign of K has to be changed in (Eq. 3.8). To summarize, depending on the parameter K, 4 reflex behaviours can be extracted:

1. **Reaching any object**: $K = k/\delta I_{ax}$, with $k > 0$,
2. **Avoid any object**: $K = k/\delta I_{ax}$, with $k < 0$,
3. **Reach conductive objects/avoid insulating objects**: $K = k/ \mid \delta I_{ax} \mid$, with $k > 0$,
4. **Reach insulating objects/avoid conductive objects**: $K = k/ \mid \delta I_{ax} \mid$, with $k < 0$.

On Fig. 3.11 an illustration of the behaviour "reaching an object" is applied when going from A to B. It has be clearly demonstrated in [5] that these behaviours can be used to navigate autonomously reaching or avoiding obstacles. It has be mentioned that all these behaviours are fully reactive and do not need any prior knowledge.

3.5.3 *Reactive Navigation Using Active Electric Sense (Memory Based)*

In the previous section we presented control strategies allowing the sensor to reach or avoid objects, here we will describe how to turn around an object based on the fact that we "recognized" it. This behaviour is based on a combination of reflex behaviours presented in the previous part. By sequentially ordering 3 reflexes we seek an object and then turns around it (see Fig. 3.11 from [13]).

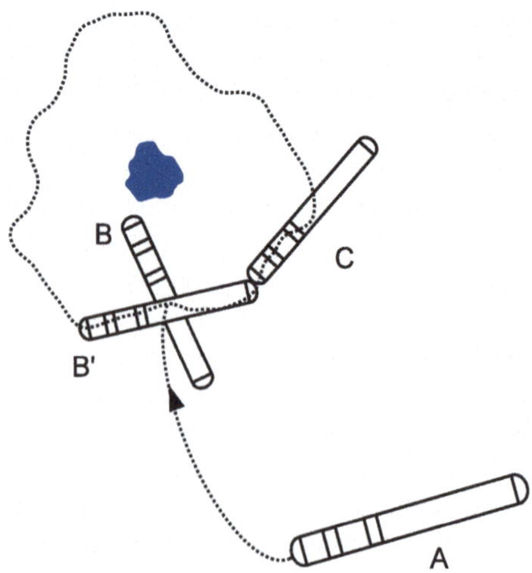

Fig. 3.11 Object exploration based on electric sense [13]

1. **Seek an object**: From A to B on Fig. 3.11 the robot is seeking an electrically non-transparent object by applying the attractive behavior with $K = k/\delta I_{ax,1}$ and $k > 0$. This behavior is maintained until $\delta I_{ax,2}$ changes its sign. The reason of this is that when getting closer to an object, more and more electric field lines are captured by the object if it is conductive. Thus, the electric field lines are concentrated on the front electrodes. On the contrary, if the object is insulating the electric field lines are pushed backward along the sensor axis. Thus, in both cases at some point $I_{ax,2}$ changes of sign.
2. **Flee from the electric influence**: From B to B' on Fig. 3.11. It corresponds to the initialization of the orbiting motion of the sensor around the object. It is obtained by applying the repulsive behavior, i.e. $K = k/ \mid \delta I_{ax,1} \mid$ and $k > 0$ until $\mid \delta I_{ax,1} \mid$ reach its minimum.
3. **Follow the boundaries**: From B' to C on Fig. 3.11. The orbiting phase is obtained by applying the law $\Omega = K(\delta I_{ax,1} - M)$. M is the value of $\delta I_{ax,1}$ measured at the last time of the previous phase.

It has to noted that the commutation between phases is ruled by events which only depend on time variation of the measurements (and not of their magnitude). Such a behaviour completely developed in [13] shows that the robot can navigate avoiding, reaching and turning around objects only using active electric sense.

3.5.4 Reach an Active Robot by Controlling a Passive Robot

We consider here a passive robot that tracks the electric lines of an electric field generated by another active robot. On Fig. 3.12a, 2 external electrodes: emitter (circle 1) and receiver (circle 2) are representing the active robot that can be considered as a docking station [4]. These electrodes are close to each other and located in one of the corners of the tank. By using the same reflex behaviours presented in the previous section we can navigate with a passive robot. This results is shown on Fig. 3.12 where the passive robot reach an emitter with 3 different starting positions [4]. The sensor's path is represented by a dotted lines and letters correspond to intermediate poses. For all position on Fig. 3.12a, the current I_{ax} is positive along the path and monotonically increases (Fig. 3.12b). On the other hand, I_{lat} converges toward its desired zero-value. The changing sign is due to the presence of the perturbative repulsive walls (from 0 to C and from 0 to B) the probe to far from the emitter first avoid the insulating walls and converge towards the emitter as it is getting closer. As shown the robot is docking on the emitter in all 3 cases with different trajectories. This experiment first presented in [4] show that navigation of a passive robot in the electric field generated by an active is also possible.

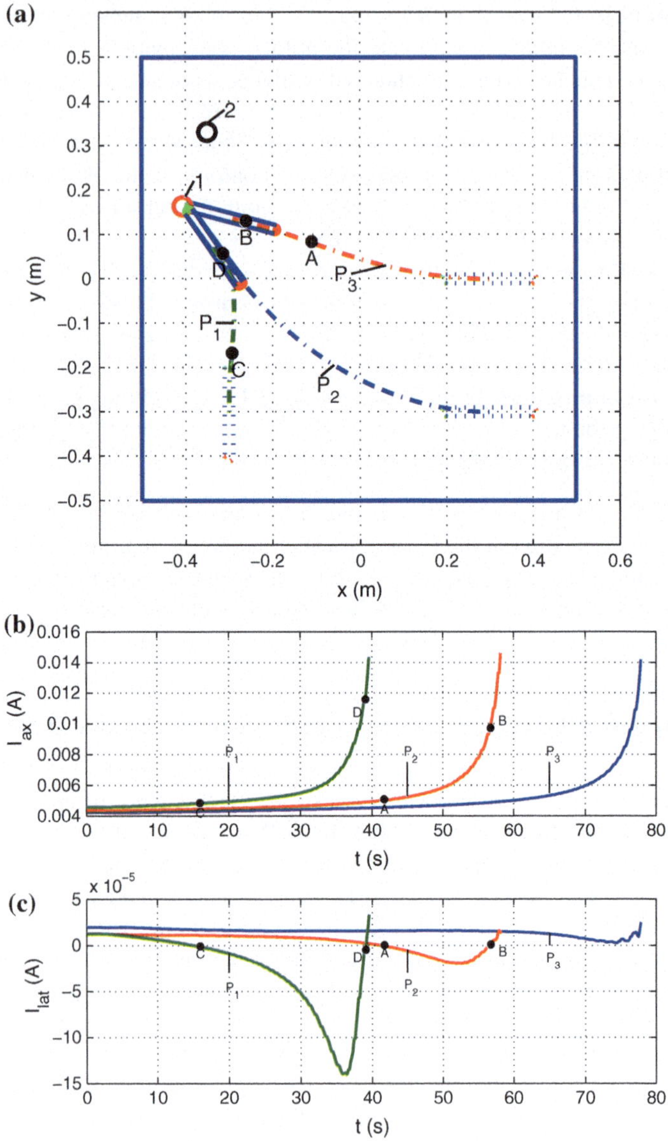

Fig. 3.12 Starting from 3 different initial poses (dashed lines), the sensor seeks the emitter until it touches it following 3 paths **a** Scene and paths P1, P2 and P3. **b** Axial currents (I_{ax}) and **c** lateral currents (δI_{lat}) for the 3 paths. A, B, C, D indicate poses where the sign of lateral currents change [4]

3.5.5 *Object Localization and Recognition from an Active Robot*

In this section we consider an active robot and an object in its range of perception than is electrically contrasted with the water (see Fig. 3.6). Our robot is moving straight alongside of the object (see Fig. 3.13) and we collect every millimeters 6 currents gathered in I_{lat}, and δI_{ax} (see Sect. 3.5.2). Our goal here is to estimate the localization, the pose and the size the object O_x that perturbates the electric field generated by the robot. To simplify the problem we will solve the problem only in 2D and we supposed that: the conductivity of the medium γ is known as well as the displacement of the robot along its trajectory in a global reference frame. We performed experiments with a conductive prolate ellipsoid (i.e. axisymmetric about its major axis), i.e., that can be described by 6 parameters: its localization x_0, y_0, its orientation θ_0, its size a, b and its electric properties (conductive or insulating) σ. Based on Sect. 3.5.2 we can already estimate the electric properties of the object σ and its side with respect to the robot as soon as the object enter in the perception range of the robot. This can be obtained looking at the sign of δI_{ax} and I_{lat}. Then, based on these information, we will estimate the missing parameters by using a greedy problem solving heuristic testing a set of potential solution in the parameter space and select the optimal solution. This method is based on the analytical model presented in [3] that models the dipolar electric response of an object immersed in a electric field produced by the robot of Fig. 3.7. Within this analytical we modeled the ellipsoidal object by its first order polarization tensor [1, 11]. Giving this analytical model, we implemented a function f that receives as parameters: a robot position (x_k, y_k, θ_k) along the trajectory $T(k)$ and an object O_x in a global reference frame. The function returns the 6 currents described in Eq. 3.7. This function that estimates 6 values we named $\widehat{I(k)}$ can be expressed as: ∀k∈ $[1, n]$ we have:

$$f(T(k), O_x) = \widehat{I(k)} = \begin{pmatrix} I_{lat}(T(k), O_x) \\ \delta I_{ax}(T(k), O_x) \end{pmatrix} \tag{3.9}$$

where k is a position along the trajectory $T(k)$ that is discretized into n positions. Finally, to estimate the ellipsoid that perturbates the electric field we compare for a set of possible object parameters the sum of the error between the real measurements and the estimated measures for each robot positions. The object with the smallest error is then chosen as the solution of our problem. This algorithm can be summarized by the Eq. 3.10.

$$\underset{0_x}{\operatorname{argmin}} \sum_{k=1}^{n} \left(\sum_{i=1}^{6} \frac{|I_i(k) - f_i(T(k), O_x)|}{|I_i(k)|} \right) \tag{3.10}$$

Figure 3.14 shows 24 experiments performed with an conductive ellipsoid which size was 33 × 16 mm. The real ellipses are displayed in red and the estimated ellipses in blue. Each results has been obtained offline using data collected by the robot (fol-

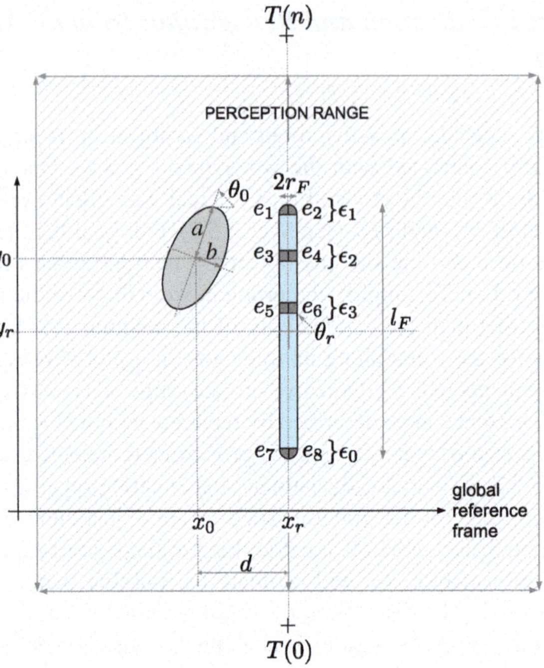

Fig. 3.13 Experimental set up with the aFish

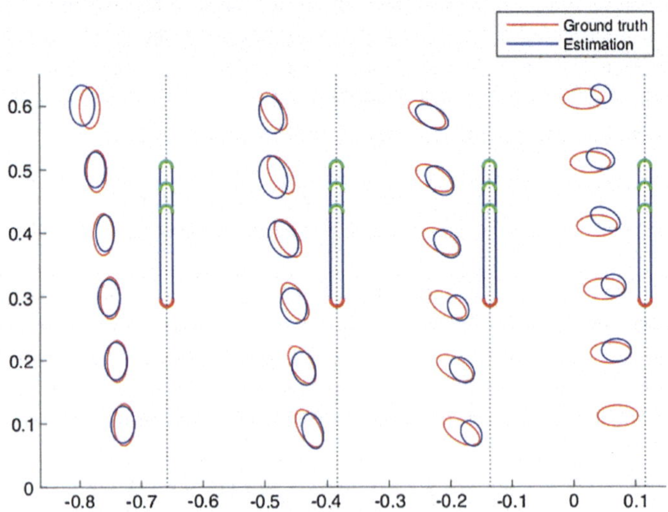

Fig. 3.14 Localization and shape estimation of a 33 × 16 conductive ellipsoid at 5 different distances. Real ellipses (red), estimates (blue)

lowing a straight line trajectory) at different 6 distances 50, 60, 70, 80, 90, 100 mm and with 4 different object orientations. As it can be seen the localization and the shape can be estimated at a short distance. As we increase the distance with the object the recognition is unfortunately getting worst as the signal noise ratio decreases. It has to be noted that for the presented results the resolution for the localization and the shape estimation was 2 mm and the angle resolution was 0.26 radians. Such a greedy method can be used to estimate our object in an acceptable time (few minutes) because all parameters are quite well constrained by the model assumption and the small range of the sensor. The position x_0, y_0 are constrained by the range of the sensor which is about the sensor length l, so $x_0 \in [-l, +l]$, and $y_0 \in [0, \pm l]$. As well, $\theta_0 \in [0, \pi]$. Then, the maximum size of the object is also known as our model consider small object (smaller than the robot length l). Then, the only parameter of the algorithm is the discretization step of the each parameters. In this last section we shown that just using electric sense we can also localize and estimate the shape of an object.

3.6 Conclusion

In this paper, we presented an overview of electric sense based perception and navigation algorithms for the underwater robots that will be used in the context of the EU project subCULTron. We presented different algorithms based on electric sense that will be used by our robots to perform reactive navigation, obstacles avoidance, objects exploration and finally object shape estimation. These methods have been illustrated by some simulation and experimental results. From these results we can conclude that in many cases electric sense can supplement common underwater sensing devices in harsh conditions that is to say confined, low light and turbid underwater environment. In the new future, this work on electric sense will be extended towards collective based behaviours and localization and mapping (SLAM).

Acknowledgements subCULTron (submarine cultures perform long-term robotic exploration of unconventional environmental niches) is supported by European Unions Horizon 2020 research and innovation program under the grant agreement no 640967. http://www.subcultron.eu.

References

1. H. Ammari, J. Garnier, H. Kang, M. Lim, S. Yu, Generalized polarization tensors for shape description. Numerische Mathematik **126**(2), 199–224 (2014)
2. Y. Bai, J. Snyder, Y. Silverman, M. Peshkin, M. MacIver, Sensing capacitance of underwater objects in bio-inspired electrosense, in *IEEE/RSJ International Conference on Intelligent Robots and Systems* (2012)
3. F. Boyer, P. Gossiaux, B. Jawad, V. Lebastard, M. Porez, Model for a sensor bio-inspired from electric fish. IEEE Trans. Robot. **28**(2), 492–505 (2012)

4. F. Boyer, V. Lebastard, C. Chevallereau, S. Mintchev, C. Stefanini, Underwater navigation based on passive electric sense: New perspectives for underwater docking. Int. J. Robot. Res. (2015)
5. F. Boyer, V. Lebastard, C. Chevallereau, N. Servagent, Underwater reflex navigation in confined environment based on electric sense. IEEE Trans. Robot. **29**(4), 945–956 (2013)
6. T. Bullock, W. Heiligenberg, *Electroreception* (Wiley, 1986)
7. A. Caputi, R. Budelli, C. Bell, The electric image in weakly electric fish: physical images of resistive objects in gnathonemus petersii. J. Exp. Biol. **201**(14), 2115–2128 (1998)
8. D. Clarke, H. Whitney, G. Sutton, D. Robert, Detection and learning of floral electric fields by bumblebees. Science **340**(6128), 66–69 (2013)
9. C. Hopkins, Electrical perception and communication, in *Encyclopedia of Neuroscience*, vol. 3, (Oxford: Academic Press, New York, 2009), pp. 813–831
10. A. Kalmijn, Electro-perception in sharks and rays. Nature **212**(5067), 1232–1233 (1966)
11. S. Lanneau, V. Lebastard, F. Boyer, Object shape recognition using electric sense and ellipsoid's polarization tensor, in *In Proceedings of the IEEE International Conference on Robotics and Automation (ICRA)*, (2016), pp. 4692–4699
12. V. Lebastard, F. Boyer, C. Chevallereau, N. Servagent, Underwater electro-navigation in the dark, in *IEEE Conference on Robotics and Automation* (2012)
13. V. Lebastard, F. Boyer, S. Lanneau, Reactive underwater object inspection based on artificial electric sense. Bioinspiration Biomim. **11**(4), 045003 (2016)
14. H.W. Lissmann, K.E. Machin, The mechanism of object location in Gymnarchus Niloticus and similar fish. J. Exp. Biol. **35**(2), 451–486 (1958)
15. B. Rasnow, The effects of simple objects on the electric field of apteronotus. J. Comp. Phys. A **3**(178), 397–411 (1996)
16. B. Rasnow, J.M. Bower, The electric organ discharges of the gymnotiform fishes. J. Comp. Phys. A **178**(3), 383–396 (1996)
17. N. Servagent, B. Jawad, S. Bouvier, F. Boyer, A. Girin, F. Gomez, V. Lebastard, P.-B. Gossiaux, Electrolocation sensors in conducting water bio-inspired by electric fish. IEEE Sens. J. **13**(5), 1865–1882 (2013)
18. G. von der Emde, S. Schwarz, L. Gomez, R. Budelli, K. Grant, Electric fish measure distance in the dark. Letters to Nature. Nature **395**, 890–894 (1998)

Chapter 4
Estimating the Trajectory of Low-Cost Autonomous Robots Using Interval Analysis: Application to the euRathlon Competition

Fabrice Le Bars, Elba Antonio, Jorge Cervantes, Carlos De La Cruz and Luc Jaulin

Abstract In this paper, we describe a method based on interval arithmetic and contractors to compute an envelope containing the trajectory of a robot from usual proprioceptive and exteroceptive data, using a simple state equation model. To illustrate the applicability of the method, data from the euRathlon 2015, a multi-domain robotics competition, will be processed to build an estimation of the trajectory of a low-cost AUV (Autonomous Underwater Vehicle), navigating with the help of acoustic communication and ranging with an ASV (Autonomous Surface Vehicle).

4.1 Introduction

In a submarine context, estimating the real trajectory of a robot after its mission is a challenging problem (see e.g. [1] for a review of possible techniques), especially for small low-cost AUVs (Autonomous Underwater Vehicles). Indeed, the GPS is not available underwater and in a small low-cost AUV, some important sensors for localization might not be available or with reduced functionality and accuracy, due to size, costs and energy restrictions:

F. Le Bars (✉) · L. Jaulin
ENSTA Bretagne (Ecole Nationale Supérieure de Techniques Avancées Bretagne), Lab-STICC/CID/PRASYS, 2 rue François Verny, 29806 Brest Cédex 09, France
e-mail: fabrice.le_bars@ensta-bretagne.fr

L. Jaulin
e-mail: lucjaulin@gmail.com

E. Antonio · J. Cervantes · C. De La Cruz
CINVESTAV (Center for Research and Advanced Studies of the National Polytechnic Institute), UMI LAFMIA CNRS 3175, 2508 Mexico City, D.F., Mexico
e-mail: eantonio@ctrl.cinvestav.mx

J. Cervantes
e-mail: jcervantes@ctrl.cinvestav.mx

C. De La Cruz
e-mail: cdelacruz@ctrl.cinvestav.mx

L. Jaulin et al. (eds.), *Marine Robotics and Applications*, Ocean Engineering & Oceanography 10, https://doi.org/10.1007/978-3-319-70724-2_4

DVL (Doppler Velocity Log): this sensor can provide the speed of the vehicle w.r.t. the sea floor or surface, which is a very important data for the localization in dead-reckoning if it is accurate enough, see e.g. [2, 3].

Sonar (SOund Navigation And Ranging): depending on its type and quality, this device can help to the vehicle localization if marks on the sea floor can be detected (see e.g. [4, 5] and [6] for a context of SLAM), if we are in a known environment (see e.g. [7, 8] and [9]) or using a bathymetry map (see e.g. [10]).

USBL (Ultra-Short BaseLine): acoustic ranging and bearing from one or several devices of known positions is a typical way to follow accurately a submarine without error accumulation, see e.g. [11].

INS (Inertial Navigation System)/AHRS (Attitude and Heading Reference System): accurate intertial and magnetic sensors to get the angles of the robot (especially the heading) are necessary to limit the error accumulation on the position estimation in a dead-reckoning context.

Additionally, few ready-to-use software suites able to fuse all those different data to build an estimation of the trajectory of a robot are available. Moreover, since most of the existing localization methods are based on probablities (see e.g. iXBlue DELPH in [12] for a Kalman-based fusion between INS, DVL, LBL, USBL, etc.), it is sometimes hard to evaluate the estimation error or detect inconsistencies. Taking into account partial information such as a position deduced by an operator that saw the robot at some time in a know place can be specially difficult to formalize and fuse with the other localization information.

For our underwater robot localization application, we propose a method based on interval arithmetic. Interval analysis has been proven to be efficient for several similar problems in robotics and other related applications:

- Localization of a robot with fugitive data that appear at some given time, see [13].
- Mosaicking, see [14].
- SLAM with indistinguishable marks, see [15].
- Getting a guaranteed approximation of the explored area, see [16].
- Detecting loops in a trajectory, see [17].
- Improvements in GPS positioning, see [18].
- Robot calibration, see [19].

In our context, we want to estimate the trajectory of a 2D robot knowing its state equations, inputs and heading at all times, with some punctual position measurements (e.g. GPS when surfacing at a waypoint) as well as range measurements with another moving robot of known position (e.g. using acoustic modems with range measurement capability). Cooperation between robots has already demonstrated its efficiency in several situations, compared to the use of a single, usually bigger and more expensive robot: see e.g. [20–22] or [23] for cooperation between 2, 3 or more AUVs, see [24] for the use of several acoustic buoys to localize underwater robots using a set-membership approach and see e.g. [25] and [26] for cooperation between an ASV and an AUV.

In this paper, we will first describe the interval-based data fusion method used, and then show its application to the trajectory estimation of the *SARDINE* AUV during the euRathlon 2015 competition (see [27] for more information about the euRathlon competition).

4.2 Set-Membership Data Fusion Method for Localization

4.2.1 Interval Arithmetic

Interval arithmetic (see e.g. [28–30] and [31]) is a numerical tool which makes it possible to solve non-linear equations and non-convex minimization problems, in a global and guaranteed way. Now, estimating the state of a robot can be seen as solving simultaneously a set of non-linear equations, for which interval analysis could be useful. This section recalls the basic notions of interval arithmetic.

An *interval* is a closed and connected subset of $\mathbb{R}$. $\mathbb{IR}$ is the set of the intervals of $\mathbb{R}$. Consider two intervals $[x]$ and $[y]$ and an operator $\diamond \in \{+, -, ., /\}$, we define $[x] \diamond [y]$ as the smallest interval which contains all feasible values for $x \diamond y$, if $x \in [x]$ and $y \in [y]$. For instance

$$\begin{aligned} [-3, 4] + [2, 5] &= [-1, 9], \\ [-1, 4] \,.\, [2, 5] &= [-5, 20], \\ [-1, 3] \,/\, [2, 5] &= [-\frac{1}{2}, \frac{3}{2}]. \end{aligned} \tag{4.1}$$

If f is an elementary function such as sin, cos, …we define $f([x])$ as the smallest interval which contains all feasible values for $f(x)$, if $x \in [x]$. A *box* or *interval vector* $[\mathbf{x}]$ is a vector whose components are intervals:

$$[\mathbf{x}] = [x_1^-, x_1^+] \times \cdots \times [x_n^-, x_n^+] = [x_1] \times \cdots \times [x_n]. \tag{4.2}$$

x^- is the interval *lower bound* and x^+ its *upper bound*. The *midpoint* (or *center*) of a bounded and non-empty interval is $\mathrm{mid}([x]) = \frac{x^- + x^+}{2}$. The *width* of a non-empty interval is defined by $w([x]) = x^+ - x^-$. By convention, $w(\emptyset) = -\infty$. If $w([x]) = 0$, $[x]$ is *degenerated*. In this case, $[x]$ is a real singleton and will be noted $\{x\}$.

For a box, the width is $w([x]) = \max\{w([x_1]), ..., w([x_n])\}$. We define also the volume of a non-empty box as

$$vol([x]) = \prod_{1 \le i \le n} w([x_i]).$$

Most of the time, if $[x]$ is an interval containing the x position of a robot, mid($[x]$) will be taken as an estimation of x, and $w([x])/2$ will be taken as the estimation error. The volume of the box $[\mathbf{x}]$ representing the 2D or 3D position estimation of the robot could be also used to evaluate the estimation precision.

4.2.2 Contraction and Propagation

ICP (Interval Constraint Propagation) is a technique (or sort of algorithm) that enables to solve efficiently (computing time is in general lower than other methods such as those involving bisections) equations on intervals. It is a combination between 2 more general notions: constraint propagation and interval arithmetic (see e.g. [32]).

Consider a constraint $\mathcal{C}$ (i.e., an equation or an inequality), some variables $x_1, x_2, \ldots$ involved in $\mathcal{C}$ and prior interval domains $[x_i]$ that contain all feasible values for the x_i's. Interval arithmetic makes possible to contract the domains $[x_i]$ without removing any feasible values for the x_i. For instance, consider the equation $x_3 = x_1 + x_2$ where the domains for x_1, x_2, x_3 are given by $[x_1] = [-\infty, 5]$, $[x_2] = [-\infty, 4]$ and $[x_3] = [6, \infty]$. These domains can be contracted to $\left[x_1'\right] = [2, 5]$, $\left[x_2'\right] = [1, 4]$ and $\left[x_3'\right] = [6, 9]$. The resulting interval calculus is as follows:

$$\begin{aligned}
x_3 = x_1 + x_2 \Rightarrow x_3 &\in [6, \infty] \cap ([-\infty, 5] + [-\infty, 4]) \\
&= [6, \infty] \cap [-\infty, 9] = [6, 9]. \\
x_1 = x_3 - x_2 \Rightarrow x_1 &\in [-\infty, 5] \cap ([6, \infty] - [-\infty, 4]) \\
&= [-\infty, 5] \cap [2, \infty] = [2, 5]. \\
x_2 = x_3 - x_1 \Rightarrow x_2 &\in [-\infty, 4] \cap ([6, \infty] - [-\infty, 5]) \\
&= [-\infty, 4] \cap [1, \infty] = [1, 4].
\end{aligned} \tag{4.3}$$

A graph of constraints can be used as a simple representation to show which variable will be used to contract the others. For example, if $z^2 = \exp(x) + y$ and $x \in [1, 4]$, $y \in [3.1, 3.2]$, $z \in [4, 8]$, then $x = \ln\left(z^2 - y\right) \Rightarrow x \in [x] \cap \ln\left(z^2 - y\right) \in [2.5, 4]$ and $z = \sqrt{\exp(x) + y} \Rightarrow z \in [z] \cap \sqrt{\exp(x) + y} \in [4, 7.6]$. Therefore, the graph for the constraint $z^2 = \exp(x) + y$ would be like in Fig. 4.1.

This contraction procedure can be performed with much more complex constraints. A contraction operator is called a *contractor* (see [33]). When more than one constraint are involved, the contractions are performed sequentially several times,

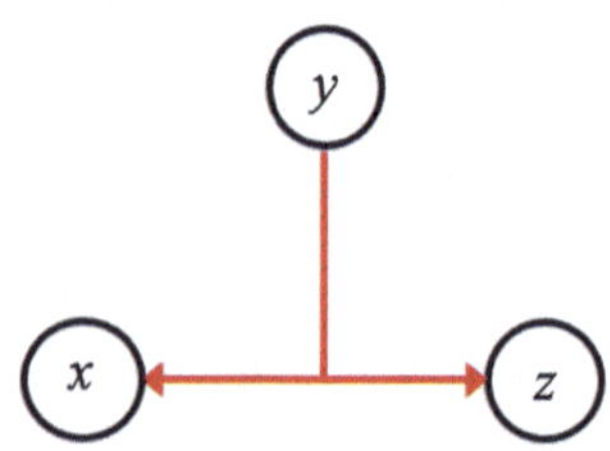

Fig. 4.1 Graph of constraints

until no more significant contractions can be observed. It can be shown that the box to which the method converges does not depend on the order to which the contractors are applied (see [28]), but the computing time is highly sensitive to this order. There is no optimal order in general, but in practice, one of the most efficient is called *forward-backward propagation*. It consists in writing the whole set of equations under the form $\mathbf{f}(\mathbf{x}) = \mathbf{y}$ where $\mathbf{x}$ and $\mathbf{y}$ correspond to quantities that can be measured (i.e., some prior interval domains are given for them). Then, using interval arithmetic, the intervals are propagated from $\mathbf{x}$ to $\mathbf{y}$ in a first step (*forward propagation*) and, in a second step, the intervals are propagated from $\mathbf{y}$ to $\mathbf{x}$ (*backward propagation*). Tools such as the IBEX library can be used to manipulate contractors, see [34].

4.2.3 Context, Assumptions and Principle of the Method Used

In this paper, we will only consider a 2D localization problem, where the vertical localization problem is considered as solved or not needed:

- Most of the submarines have accurate depth and/or altimeter sensors compared to x, y localization sensors.
- Due to the submarine depth control at low depth and the fact that the acoustic modem on the surface vehicle is 1 m below the surface, the submarine acoustic modem can be considered in the same plane as the surface robot acoustic modem.
- The robot's trajectory is most of the time in the same plane, the depth changes are typically almost punctual: submerging in the beginning to reach target mission depth, surfacing at the end or to get GPS data.

We will also assume that there are no outliers in the data: they were manually removed, e.g. obvious temporary bad GPS fixes at some surfacing points (GPS fixes will be considered as punctual: for each fix, only 1 position at a specific time has been taken into account), and bad range measurements or data communications were automatically rejected. If outliers cannot be removed manually, a method using relaxed intersection is proposed in [7].

The method used in this paper is inspired from [4] and [5]. In our set-membership context, we can describe our problem as follows

$$\begin{cases} \dot{\mathbf{x}} = \mathbf{f}(\mathbf{x}, \mathbf{u}) & \text{(evolution equation)} \\ \mathbf{y} = \mathbf{g}(\mathbf{x}) & \text{(observation equation)} \\ r_i = r(\mathbf{x}, \mathbf{m}) & \text{(ranging equation)} \end{cases} \tag{4.4}$$

where $\mathbf{x}(t)$ is the state vector of the submarine, $\mathbf{u}(t)$ is its input vector, $\mathbf{y}(t)$ is its output vector, $\mathbf{m}$ is the location of the surface robot and $\mathbf{r}_i$ is the ith range measurement between the 2 robots. We consider that

- At some time t, we have a box $[\mathbf{x}](t)$ containing the state vector:

$$\mathbf{x}(t) \in [\mathbf{x}](t) \tag{4.5}$$

- For all $t \in [t_0, t_f]$, we have boxes enclosing the $\mathbf{u}(t)$ and $\mathbf{y}(t)$:

$$\forall t \in [t_0, t_f], \mathbf{u}(t) \in [\mathbf{u}](t) \text{ and } \mathbf{y}(t) \in [\mathbf{y}](t) \tag{4.6}$$

- We have a finite subset $\mathcal{M} \subset [t_0, t_f]$ such that

$$r_i(t) \in [r_i](t). \tag{4.7}$$

If $t \in \mathcal{M}$, the ranging measurement i has been made at time t and we have a box enclosing $\mathbf{m}(t)$.

For simplicity, we assume that the submarine motion can be described by the following evolution equation (inspired from [7]):

$$\begin{cases} \dot{x} = v\cos\theta \\ \dot{y} = v\sin\theta \\ \dot{\theta} = u_2 - u_1 \\ \dot{v} = (u_1 + u_2) - v \end{cases} \tag{4.8}$$

where (x, y) are the 2D coordinates of the robot, θ is its heading, v its speed, u_1 the input for the right thruster, u_2 the input for the left thruster (see Fig. 4.2).

This equation is similar to the classical tank/wheelchair model (see e.g. [35]), with very simple adaptations to simulate the fluid friction effects. We have chosen here a normalized model with the friction coefficient (in front of v) equal to 1, but in practice, the coefficient is different and needs to be estimated (e.g. empirically).

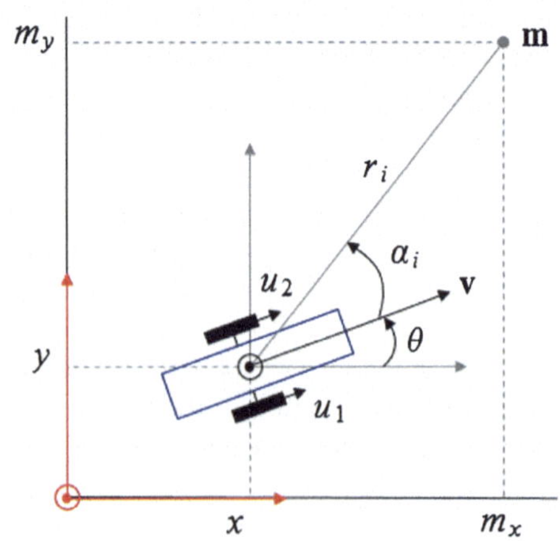

Fig. 4.2 Description of the submarine variables

The observation equation is

$$\mathbf{y} = \begin{pmatrix} x \\ y \\ \theta \end{pmatrix} \tag{4.9}$$

because (x, y) can be measured sometimes by the GPS when the robot is near the surface and θ is measured by the AHRS at all times.

The ranging equation for the ith communication and range measurement is

$$r_i = \sqrt{(x - m_x)^2 + (y - m_y)^2} \tag{4.10}$$

where r_i is the distance between the submarine and the surface robot (measured by the internal ranging function of the acoustic modems) and $\mathbf{m} = (m_x, m_y)$ is the position of the surface robot (measured by its GPS and sent to the submarine through the acoustic modems). It is worth noting that to be accurate in practice, due to the communication and ranging time, we should consider the fact that the position of the robots might have changed between the moment when the range measurement is made, the moment when the position of the surface vehicle is measured, and the moment when it is received by the submarine. We could also have clock synchronization errors between robots (see e.g. [36] for methods to handle this). In our case, those uncertainties will be included inside the uncertainty for $\mathbf{m}$ and r_i.

As a consequence, our problem can be written as follows:

$$\begin{cases} \dot{x} &= v \cos\theta \\ \dot{y} &= v \sin\theta \\ \dot{\theta} &= u_2 - u_1 \\ \dot{v} &= (u_1 + u_2) - v \\ r_i &= \sqrt{(x - m_x)^2 + (y - m_y)^2} \end{cases} . \tag{4.11}$$

To solve this problem, we will use the ICP techniques described previously:

- First we need to use the observation equation to contract the known GPS positions of the robot, see Fig. 4.3.
- Then, a forward-backward propagation w.r.t. time on the differential equations discretized using e.g. Euler method can be used to get a first evaluation of the trajectory, see Fig. 4.4.
- Contracting the ranging equation should improve the position estimation of the robot at the specific ranging times, see Fig. 4.5.
- Repeating all the previous contraction operations until no more significant improvement on the trajectory estimation should propagate the punctual contractions on the whole trajectory.

Note that if needed, we could take into account more general exteroceptive data such as bearing measurements, see e.g. [37] or [38] for related work.

Fig. 4.3 Graph of constraints for the observation equation

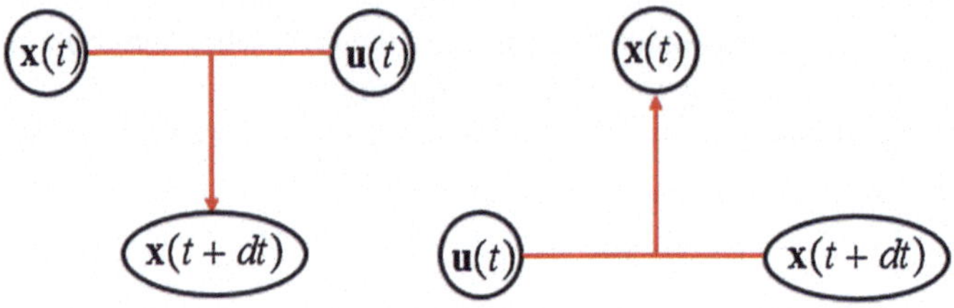

Fig. 4.4 Graph of constraints for the evolution equation (left: forward with time, right: backward with time)

Fig. 4.5 Graph of constraints for the ranging equation

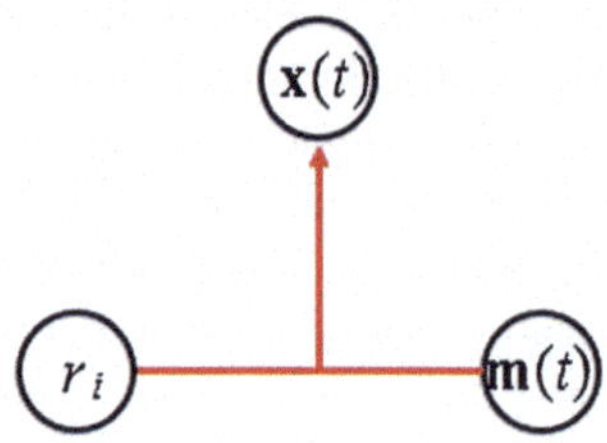

4.3 Application

4.3.1 Description of the Experiment

The euRathlon 2015 competition (see [27]) proposed a situation where marine and submarine robots had to collaborate with ground and aerial robots to analyze the state of an area after a possible earthquake and determine which pipeline from a mock nuclear plant should be stopped to stem a leak while letting other pipelines cool the reactor. The submarine data presented in this paper corresponds to the AUV navigating through the pipelines structures area, and searching for the right one (2 groups of pipelines separated by 50 m and at a depth of around 5 m were to be inspected, with colored objects indicating their state). Meanwhile, aerial and ground robots were inspecting the damages around and inside the mock nuclear plant.

We must emphasize that these data were taken during a competition, therefore they have not been specifically done to illustrate the method described here. This implies that they are not perfect, for example: compasses of the robots were not calibrated correctly, the competition area was probably not deep enough to get a good acoustic communication, the acoustic modem on the ASV only started in the middle of the experiment, etc. This is to show that the approach can be efficient on real-world data.

Fig. 4.6 AUV and ASV

The *SARDINE* AUV was equipped with a pressure sensor to get his depth, a mid-cost AHRS to get its heading, a GPS to get its position near the surface, a rotating sonar (not used for this part of the experiment), 2 cameras (not used for this part of the experiment) and a low-cost acoustic modem with ranging capability. In particular, it did not have a DVL or other sensor to evaluate its speed. It is controlled in the plane using 2 horizontal thrusters, and a central vertical thruster controls his depth (it is statically equilibrated to have an almost neutral buoyancy without roll or pitch), see Fig. 4.6. The ASV was controlled using 2 thrusters and was equipped with a GPS, a low-cost AHRS, a camera (not used for this part of the experiment) and the same acoustic modem as the AUV, installed at 1 m below the hull to avoid immediate reflexion of acoustic waves on it.

During the experiment, the robots had their own embedded position estimation algorithms. The principle of the algorithm to estimate the position used inside the submarine was similar to what is described in this paper, but with the following differences:

- No backward propagation w.r.t. time.
- Different coefficients were used in the state equations (few time was spent to adjust them during the experiment).
- In case of inconsistencies between data, specific non-optimal fallback choices were made.

It is also important to note that this estimation was not always used by the embedded control part, see Fig. 4.7:

- In the beginning yes.
- However for the "spirals", no, an open loop was used.

Here is a description of the trajectory (in Fig. 4.7) evaluated by the embedded algorithm and retrieved just after the experiment, without further processing (we will refer to this estimation as the "realtime estimation" in what follows):

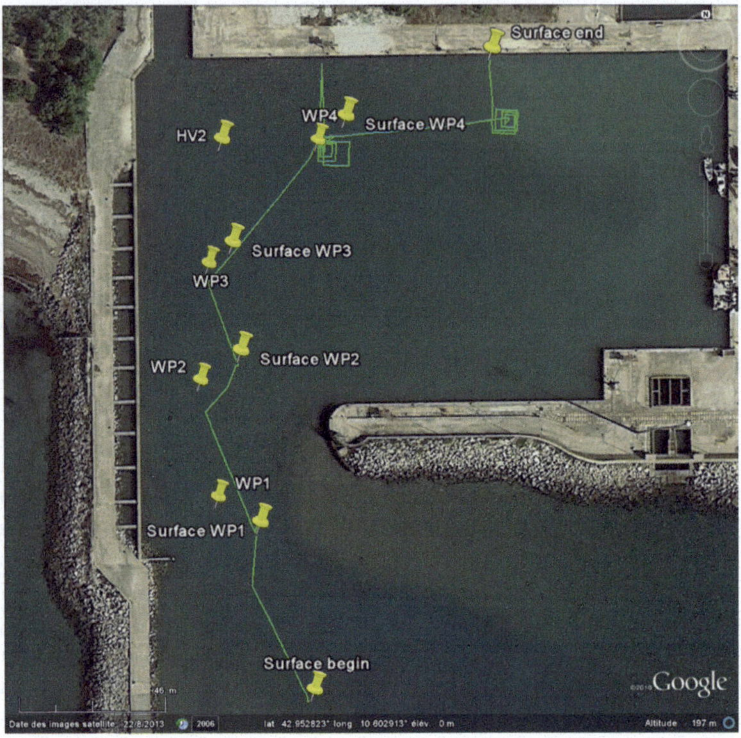

Fig. 4.7 Realtime trajectory estimation retrieved from the robot just after the experiment. Note that we do not have the ground truth for this experiment and that in the beginning, the robot is following waypoints ("WP1"..."WP4"), while in the end, it is making 2 spirals without following specific waypoints

- The ASV was always staying around the same waypoint "HV2".
- The AUV submerged at "Surface begin" point and started to follow the waypoints, surfacing at each one ("WP1", "WP2", "WP3", "WP4"). During this phase, we clearly see the "jumps" in the trajectory estimation due to the GPS when surfacing: between 2 waypoints, the trajectory estimation appears to be perfectly in the direction of the target waypoints, however this estimation is corrected when the robot surfaces and gets the GPS. When at the surface, we also see additional "jumps" due to the fact that the GPS provides very noisy data when it is at the limit to not be able to see the satellites.
- After "WP4", it started to make a spiral around, using relative control commands (i.e. without trying to use its absolute position estimation). The trajectory estimation has some "jumps" at this point also due to the corrections made with the range measurements.
- Then, it moved to the East and made another spiral.
- Finally it surfaced at the end of the mission. The total distance covered by the robot was approximately 500 m.

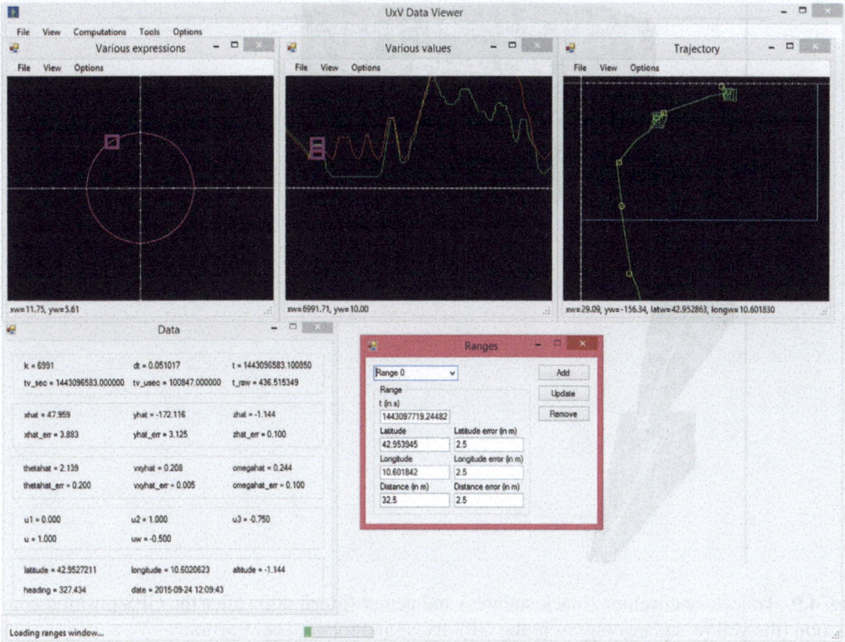

Fig. 4.8 UxV Data Viewer, a software made to post-process the data of robots using interval methods

4.3.2 Results

A program available on

`www.ensta-bretagne.fr/lebars/sardine2015/`

has been made to make the offline processing of the data, see Fig. 4.8.

All the state variables are first initialized with intervals centered on the realtime estimation and with an infinite width, while the inputs are loaded as degenerated intervals. After loading all the data as intervals, the first operation that should be done is to contract the data w.r.t. the measured punctual GPS positions (using the observation equation). To propagate the effect of those punctual known positions, we need then to use the evolution equation of the submarine to contract forward in time (see Fig. 4.9), and then backward (see Fig. 4.10).

For the forward contraction, the x and y errors (see Fig. 4.11) increase linearly over the time. Between e.g. the second and the third waypoint, we clearly see 3 different inclinations of the x and y error curves:

- The first is when the robot is stopped at the surface or going down after the GPS fix, but is not trying to move in the horizontal plane.

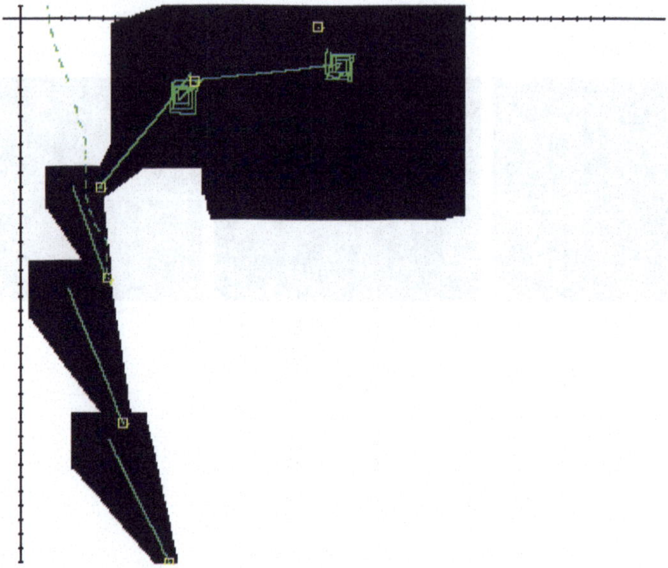

Fig. 4.9 Trajectory envelope (black squares) and center (green dots) after the GPS position contraction (the yellow squares represent the GPS fixes) and forward contraction

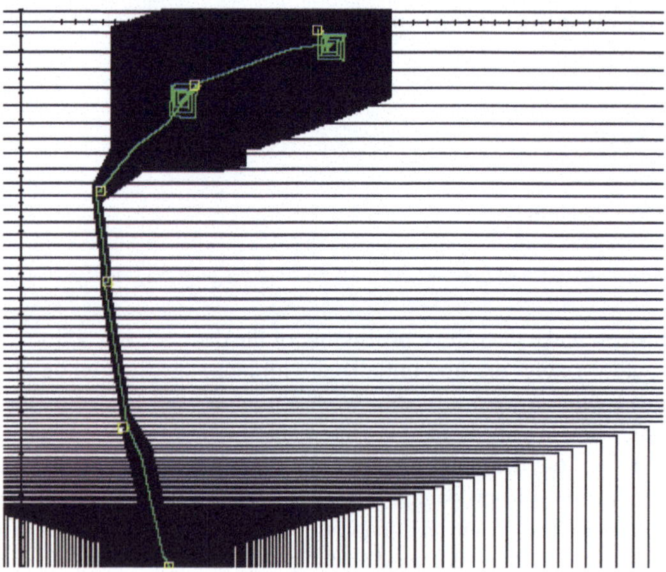

Fig. 4.10 Trajectory envelope (black squares) and center (green dots) after the GPS position contraction (the yellow squares represent the GPS fixes), forward contraction and backward contraction

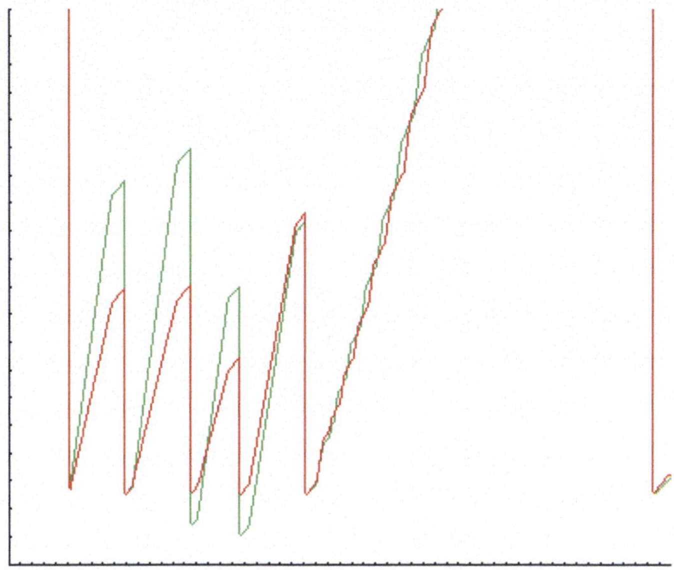

Fig. 4.11 Trajectory error w.r.t. time (x: green, y: red) after the GPS position contraction and forward contraction. Axis are [0, 3600] s × [0, 20] m. The graphic is cut on the top to be able to see in the same time the general trends and the specific parts discussed in this section

- The second when the robot is moving from one waypoint to the next.
- The third is similar to the first: the robot is surfacing but is not trying to move in the horizontal plane.

If we were only doing the backward contraction (i.e. without forward contraction before), the errors would decrease linearly.

After the forward-backward contractions, we notice a step in the middle of the y error graphics, between e.g. the first and the second waypoint (see Fig. 4.12). This is due to the fact that the robot seems to have a bias (bad magnetic calibration, currents, non symmetric parts on the robot could explain this) and the GPS position measured at the surface was a surprise w.r.t. the expected surfacing position. This is consistent with the small width of the boxes (i.e. small width of $[x]$) at this part of the trajectory: the GPS is about to be inconsistent with the state equations on the x coordinate, but it is still consistent enough to not generate an empty set. To be more clear, we can take a simple example where a car travels between 2 cities in a straight line (1 dimension problem) at a limited speed in a known time. If:

- The speed limit is 110 km/h (the car can be slower, but not faster),
- The time of travel was 3 h 10 min,
- The distance between the cities is 330 km,

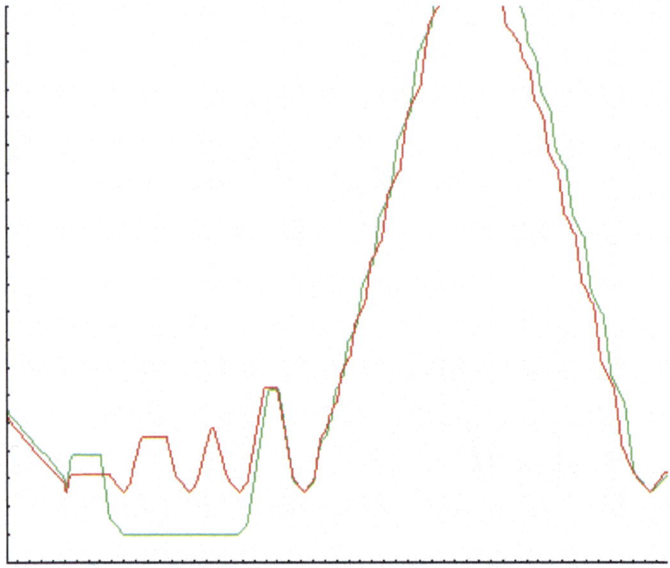

Fig. 4.12 Trajectory error w.r.t. time (x: green, y: red) after the GPS position contraction, forward contraction and backward contraction. Axis are [0, 3600] s × [0, 20] m. The graphic is cut on the top to be able to see in the same time the general trends and the specific parts discussed in this section

We can estimate the position of the car at all times (except at the very beginning and the very end where we know in which city we are) with the same error, which corresponds to the distance covered in 10 min at 110 km/h, i.e. 18.33 km.

If we continue the analysis of Fig. 4.12, there is another case where there is almost no step (triangular curve shape): between the second and the third waypoint. In this case, the robot was always in the middle of the box, there was no surprise at the end. In Fig. 4.10, we can also mention that there are very big boxes in the beginning before the first GPS fix, and the dotted green part of the trajectory visible in Fig. 4.9 corresponds also to the center of some of them before the backward contraction.

If we now use the ranging equation to contract the trajectory, and then make again a forward and backward contraction to propagate the effect of the punctual range measurements, we get the Figs. 4.13 and 4.14. Since the range measurements were made mainly when the submarine was far on the right of the surface robot, only the x coordinate estimation is significantly improved.

The final post-processed trajectory estimation is compared with the realtime estimation in Fig. 4.15. We clearly see that for the final "spiral", the robot was not aware that it drifted closer to the North wall than expected. Indeed, the operators following the robot during the competition (a small buoy at the surface was attached to the submarine as a safety mark) confirmed that the robot was very likely to be close to the wall.

Fig. 4.13 Final trajectory envelope (black squares) and center (green dots), with GPS fixes (yellow squares)

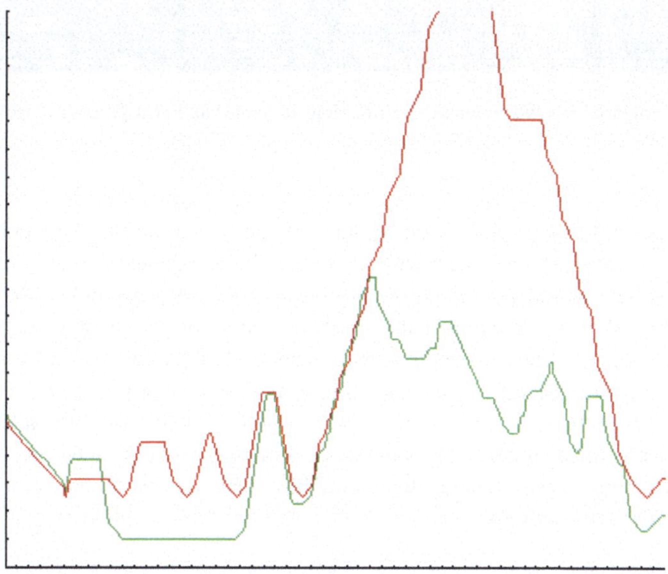

Fig. 4.14 Final trajectory error w.r.t. time (x: green, y: red). Axis are [0, 3600] s × [0, 20] m. The graphic is cut on the top to be able to see in the same time the general trends and the specific parts discussed in this section

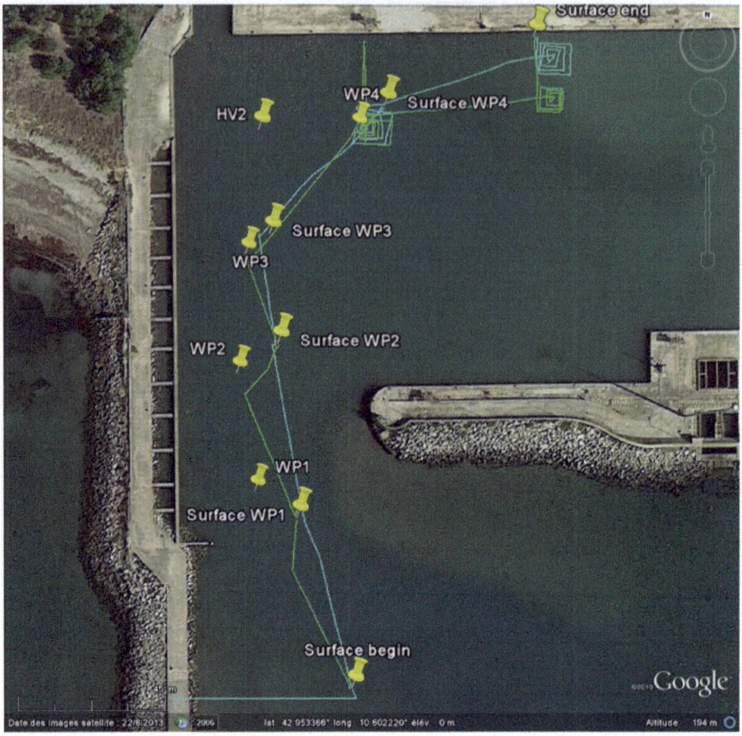

Fig. 4.15 Comparison between realtime estimation (in green) and post-processed estimation (in blue)

It is important to note that in the context of such a submarine competition, it is very difficult (for example necessary sensors would be expensive, would perturbate other sensors or would not fit on all the robots), to get a ground truth. Most of the time the robot is not even visible from the operators point of view when the robot is navigating in the sea. In those circumstancies, a guaranteed set-membership method enables to quickly obtain an estimation of the robot trajectory despite the non-linearities of the equations and if all the assumptions made are valid, it guaranties that the robot is inside the trajectory envelope, otherwise it helps detecting unknown events generating inconsistencies that would have been missed otherwise if we were using probabilistic methods.

4.4 Conclusion

In this paper, we described a set-membership method used to take into account different types of proprioceptive and exteroceptive data to compute the envelope of the trajectory of a mobile robot. The method has been applied to the localization of a submarine with puntual GPS positions when at the surface and punctual range measurements with a surface vehicle of known position. A post-processing software has been made to demonstrate the simple use of interval methods in this kind of localization problem, and a similar algorithm was run in realtime on the submarine (with some limitations such as the absence of backward propagation in time because the robot does not know its future GPS fixes). The method is general enough to be also used for other kind of robots.

References

1. L. Paull, S. Saeedi, M. Seto, H. Li, AUV navigation and localization: a review. IEEE J. Ocean. Eng. **39**(1), 131–149 (2014)
2. S. Hou, S. Peng, Z. Yan, W. Zhang, Research on the error model of INS/DVL system for autonomous underwater vehicle, in *IEEE International Conference on Automation and Logistics, China* (2008)
3. C. Vrignaud, J. Meyrat, *Use of a DVL in an Autonomous Underwater Vehicle for a Rapid Environmental Assessment* (In ADCP in Action, San Diego, 2009)
4. L. Jaulin, A nonlinear set-membership approach for the localization and map building of an underwater robot using interval constraint propagation. IEEE Trans. Robot. **25**(1), 88–98 (2009)
5. F. Le Bars, A. Bertholom, J. Sliwka, L. Jaulin, Interval SLAM for underwater robots; a new experiment, in *NOLCOS 2010, Bologna, Italy* (2010)
6. D. Ribas, P. Ridao, J. Neira, J.D. Tardos, SLAM using an imaging sonar for partially structured underwater environments, in *2006 IEEE/RSJ International Conference on Intelligent Robots and Systems* (2006)
7. L. Jaulin, Robust set-membership state estimation; application to underwater robotics. Automatica **45**(1), 202–206 (2009)
8. D. Ribas, J. Neira, P. Ridao, J.D. Tardos, AUV localization in structured underwater environments using an a priori map, in *7th IFAC Conference on Manoeuvring and Control of Marine Crafts, Lisboa, Portugal* (2006)
9. F. Maurelli, Y. Petillot, A. Mallios, P. Ridao, S. Krupinski, Sonar-based AUV localization using an improved particle filter approach, in *OCEANS 2009* (2009)
10. N. Fairfield, D. Wettergreen, Active localization on the ocean floor with multibeam sonar, in *OCEANS 2008* (2008)
11. P. Rigby, O. Pizarro, S.B. Williams, Towards geo-referenced AUV navigation through fusion of USBL and DVL measurement. Oceans J., 1–6 (2006)
12. R. Girault, P. Alain, Advances in geophysical sensor data acquisition, in *12th ASF Congress, Rennes, France* (2009)
13. F. Le Bars, J. Sliwka, O. Reynet, L. Jaulin, State estimation with fleeting data. Automatica **48**(2), 381–387 (2012)
14. M. Laranjeira, L. Jaulin, S. Tauvry, Building underwater mosaics using navigation data and feature extraction. Reliab. Comput. (2016)
15. L. Jaulin, Pure range-only SLAM with indistinguishable marks. Constraints (2016)
16. B. Desrochers, L. Jaulin, Computing a guaranteed approximation the zone explored by a robot. IEEE Trans. Autom. Control (2016)
17. C. Aubry, R. Desmare, L. Jaulin, Loop detection of mobile robots using interval analysis. Automatica (2013)
18. V. Drevelle, P. Bonnifait, Interval-based fast fault detection and identification applied to radionavigation multipath. Int. J. Adapt. Control Sig. Process. **30**, 154–172 (2016)
19. J.A. Dit Sandretto, G. Trombettoni, D. Daney, G. Chabert, Certified calibration of a cable-driven robot using interval contractor programming, in *Computational Kinematics*, eds. by F.

Thomas, A. Perez Gracia. Mechanisms and Machine Science, vol. 15 (Springer, Netherlands, 2014), pp. 209–217

20. B. Allotta, L. Pugi, R. Costanzi, G. Vettori, Localization algorithm for a fleet of three AUVs by INS, DVL and range measurements, in *15th International Conference on Advanced Robotics (ICAR)* (2011)
21. J. González, I. Masmitjà, S. Gomáriza, E. Molino, J. del Río, A. Mànuel, J. Busquets, A. Guerrero, F. López, M. Carreras, D. Ribas, A. Carrera, C. Candela, P. Ridao, J. Sousa, P. Calado, J. Pinto, A. Sousa, R. Martins, D. Borrajo, A. Olaya, B. Garau, I. González, S. Torres, K. Rajan, M. McCann, J. Gilabert, AUV based multi-vehicle collaboration: Salinity studies in Mar Menor coastal lagoon. IFAC Proc. Vol. **45**(5), 287–292 (2012)
22. G. Rui, M. Chitre, Cooperative positioning using range-only measurements between two AUVs, in *OCEANS 2010* (2010)
23. M. L'Hour, V. Creuze, French archaeology's long march to the deep-the lune project: building the underwater archaeology of the future, in *ISER: International Symposium on Experimental Robotics*, vol. 109, The 14th International Symposium on Experimental Robotics, ed. by M. Ani Hsieh, O. Khatib, V. Kumar (Springer International Publishing, Marrakech/Essaouira, Morocco, 2014), pp. 911–927
24. A. Caiti, A. Garulli, F. Livide, D. Prattichizzo, Localization of autonomous underwater vehicles by floating acoustic buoys: a set-membership approach. IEEE J. Ocean. Eng. **30**(1), 140–152 (2005)
25. J. Melo, A. Matos, Guidance and control of an ASV in AUV tracking operations, in *OCEANS 2008* (2008)
26. M.F. Fallon, G. Papadopoulos, J.J. Leonard, *Cooperative AUV Navigation Using a Single Surface Craft* (Springer, Berlin, 2010), pp. 331–340
27. G. Ferri, F. Ferreira, V. Djapic, Y. Petillot, M. Palau, A. Winfield, The euRathlon 2015 grand challenge: the first outdoor multi-domain search and rescue robotics competition-a marine perspective. Mar. Technol. Soc. J. **50**(4), 81–97 (2016)
28. L. Jaulin, M. Kieffer, O. Didrit, E. Walter, *Applied Interval Analysis, with Examples in Parameter and State Estimation, Robust Control and Robotics* (Springer, London, 2001)
29. R.B. Kearfott, V. Kreinovich (eds.), *Applications of Interval Computations* (Kluwer, Dordrecht, 1996)
30. R.E. Moore, *Methods and Applications of Interval Analysis* (SIAM, Philadelphia, 1979)
31. L. Jaulin, J. Ninin, O. Reynet, IAMOOC : a MOOC on interval analysis with applications to parameter estimation and robot localization (2015), http://iamooc.ensta-bretagne.fr/
32. L. Jaulin, E. Walter, Guaranteed robust nonlinear minimax estimation. IEEE Trans. Autom. Control **47**(11), 1857–1864 (2002)
33. G. Chabert, L. Jaulin, QUIMPER, a language for quick interval modelling and programming in a bounded-error context. Artif. Intell. **173**, 1079–1100 (2009)
34. G. Chabert, *IBEX 2.0. Ecole des Mines de Nantes* (2013), http://www.emn.fr/z-info/ibex/
35. L. Jaulin, *Mobile Robotics*, ISTE edn. (2015)
36. A. Bethencourt, L. Jaulin, Solving non-linear constraint satisfaction problems involving time-dependant functions. Math. Comput. Sci. **8**(3) (2014)
37. B. Desrochers, L. Jaulin, A minimal contractor for the polar equation; application to robot localization. Eng. Appl. Artif. Intell. (2016)
38. M.S. Ibn Seddik, L. Jaulin, J. Grimsdale, Phase based localization for underwater vehicles using interval analysis. Math. Comput. Sci. **8**(3) (2014)

Chapter 5
Marine Robots in Environmental Surveys: Current Developments at ISME—Localisation and Navigation

Andrea Caiti, Riccardo Costanzi, Davide Fenucci, Benedetto Allotta, Francesco Fanelli, Niccolò Monni and Alessandro Ridolfi

Abstract Despite the growing interest that marine engineering has received during the past few decades, autonomous underwater navigation can be still considered a daunting task. Unfavourable environmental conditions and limitations on the typologies of available sensors increase the difficulties autonomous vehicles encounter during the execution of planned missions. As of today, one of the main challenges researchers face is the precise localisation of underwater vehicles, where the limitation of the error drift over time becomes extremely important for long navigation missions. In recent years, the authors, within ISME—Interuniversity Center of Integrated Systems for the Marine Environment, extensively worked on autonomous underwater navigation, with special focus on positioning techniques. Working in parallel, research was conducted on the topics of both the improvement of state-of-the-art navigation techniques and on the employment of local sensor networks to periodically reset position errors. This contribution reports the most significative results obtained by the authors during these years.

All the Authors participate in the Italian Inter-University Research Centre on Integrated Systems for the Marine Environment (ISME)—www.isme.unige.it.

A. Caiti (✉) · R. Costanzi · D. Fenucci
Dipartimento di Ingegneria dell'Informazione and Research Center "E. Piaggio", University of Pisa, Largo Lucio Lazzarino 1, 56122 Pisa, Italy
e-mail: andrea.caiti@unipi.it

B. Allotta · F. Fanelli · N. Monni · A. Ridolfi
Department of Industrial Engineering, University of Florence, Via di Santa Marta 3, 50139 Florence, Italy
e-mail: benedetto.allotta@unifi.it

L. Jaulin et al. (eds.), *Marine Robotics and Applications*, Ocean Engineering & Oceanography 10, https://doi.org/10.1007/978-3-319-70724-2_5

5.1 Introduction

As of today, autonomous underwater navigation remains a challenging task. Indeed, the number of available sensors for accurate localisation is limited by the physics of the marine environment, hence Autonomous Underwater Vehicles (AUVs) often rely on inertial and velocity sensors in order to obtain an estimate of their pose either through dead reckoning or exploiting more sophisticated navigation algorithms (such as Kalman filters and their extensions [1]).

Historically, the first AUVs were employed in the military field; nowadays, such vehicles can be considered a cost affordable solutions for their use in many applications in the industry or in the civilian field: only to cite some examples, they are employed in the Oil&Gas sector to carry out high-depth tasks in environments unsuitable for divers. They find also application in the field of underwater geology, biology and archaeology, where their optical or acoustic payload is employed for data collection and detection or monitoring activities of targets of interests on the seabed.

Regardless of the particular field of application, due to the high level of accuracy required for the execution of the above-mentioned tasks, it is crucial that AUVs exploit precise self-localisation algorithms such as those mentioned at the beginning of this Section [2].

In addition to task execution quality, a high performance degree in an AUV navigation systems is necessary for a suitable geo-referencing process of the payload data collected during a mission. For long lasting missions, e.g. required for some hydrography or geophysics applications, this is particularly true as the navigation error is subjected to an unavoidable drift. The two ways to maintain the error limited are on one side to reduce the velocity of error growth by working on the navigation filter, on the other side systems for AUV positioning based on local static or dynamic sensor networks are a fundamental tool for a periodic reset of the error without the necessity of dedicated procedures as e.g. surfacing to get the Global Positioning System (GPS) signal.

This work focuses on the research activity carried out by the authors, within ISME—Interuniversity Center of Integrated Systems for the Marine Environment, on both the two topics with two dedicated sections.

In the recent past, the authors extensively worked on the topics of autonomous navigation: in particular, they focused on pose estimation algorithms for autonomous underwater vehicles. Both position and attitude estimation have been considered and independently addressed without introducing considerable errors as the dynamics of underwater vehicles is non-fast. This way, different strategies based on state-of-the-art approaches for the two different problems were chosen as a starting point for the research. For what it concerns attitude estimation, the authors worked on a nonlinear complementary filter based on a standard solution [3]; several modifications were applied to the algorithm there proposed in order to better adapt it to the underwater field of application. The resulting algorithm, fusing data coming from a low-cost off-the-shelf Micro-Electro-Mechanical Systems (MEMS) Inertial Measurement Unit

(IMU) and a more accurate single-axis Fibre Optic Gyroscope (FOG), is able to accurately compute an attitude estimation even under the presence of unknown magnetic disturbances, which constitute a severe impairment for autonomous vehicles which are required to navigate, e.g., close to a particular seabed (from a geological point of view), or to modern metal wrecks [4, 5].

Regarding position estimation, instead, the authors resorted to an Unscented Kalman Filter (UKF) based solution [6] exploiting a suitably developed mixed kinematic and dynamic model of an underwater vehicle [1, 7]. In contrast with Extended Kalman Filter (EKF) based algorithms [8], which are frequently proposed in the literature, the UKF does not require the computation of the derivatives of the used vehicle models, hence allowing its use even in the case of nonlinear or discontinuous models (such as usual AUVs motion models). After an initial offline validation of the proposed UKF-based filter, the authors recently validated it online during suitable sea tests.

Recently, the authors focused on simultaneous estimation of AUV pose and of sea currents acting on the vehicle itself. Such estimate is computed through a suitable modification of the previously adopted system model, adapting it in order to include current contribution to the vehicle dynamics. Indeed, the online estimation of sea currents is very important for autonomous navigation because (a) it allows to obtain information about surrounding sea conditions without resorting to expensive current sensors which have to be placed in a fixed position, (b) it improves position estimation accuracy, compensating for motion caused by external disturbances (unknown system inputs) and not by thruster action (known system inputs), and (c) it yields the possibility to intelligently plan autonomous missions, advantageously exploiting the acquired information to reduce on board power consumption.

For what it concerns the topic of acoustic sensor networks for vehicle positioning they are based on suitable acoustic modems that in addition to their communication functionalities allow to determine a reliable estimate of the vehicle position [9, 10]. These sensors, such as Long BaseLine (LBL) and Ultra Short BaseLine (USBL) can be even employed over long distances (some kilometres), thanks to underwater sound propagation properties. For instance, the USBL localisation system is based on a pair of transducers, one mounted on the vehicle and the other one in a measurable location (e.g. a support vessel). Usually, the USBL sensor is mounted under the support vessel, whose position is measurable by means of GPS, and it is used to emit acoustic signals addressed to the vehicle-mounted transducer, whose response is then elaborated in order to obtain a measure of the relative pose. Additionally, USBL systems can be found in the *inverted* configuration (iUSBL), where the transducer mounted on the vehicle is used to have a measurement of relative position directly on board the AUV.

About estimation error reset based on position measurements, two successful examples of acoustic-aided navigation can be found in [11], where a dynamic acoustic sensor network is used for cooperative navigation purposes with a vehicle equipped with a USBL transducer, and in [12], where the authors show how a vehicle-mounted USBL transducer can perform Simultaneous Localisation and Mapping (SLAM) in an area where a series of acoustic modems, used as SLAM landmarks, are moored. Good results are obtained both in the case of a priori knowledge about the number of acoustic nodes and in the case of completely unknown network, improving the localisation performance of a vehicle equipped with a GPS receiver and a cheap MEMS IMU only.

This contribution is divided into two parts: the first part is focused on position and sea current estimation; particular attention is given to the equations used to model the kinematics and dynamics of the vehicle, including the propulsion model, and the interaction between marine current and the vehicle itself. Then, simulation results of current estimation, based on experimental data, are reported. The second part of the contribution, on the other hand, describes the two methods for bounding the navigation estimation error based on acoustic measurements. Their possible fields of application are different, but a common feature is that both of them do not require time and money consuming pre-mission calibration procedures aimed at determining the absolute position of the network nodes.

In the conclusive section a summary of the strength points of the proposed methods is proposed along with some considerations on how the different approaches, till the moment singularly tested and/or validated, can be adopted simultaneously to obtain a sum of benefits in terms of both error growth velocity and error value bound.

5.2 Simultaneous AUV Navigation State and Sea Current Estimation Using UKF

This Section is focused on the use of an augmented-state Unscented Kalman Filter for online estimation of the position of an AUV and of sea currents acting on it exploiting only information acquired with on-board sensors, without resorting to expensive external sensors. At first, a mixed kinematic and dynamic model of an AUV is used, including a model for its propulsion system; then, it is shown the suitable modification applied to such model due to the introduction of sea currents, i.e. the augmentation of the system state in order to include the fixed-frame components of the current itself. Finally, offline simulation results based on experimental data collected at sea are given.

5.2.1 AUV Modeling

The classic formulation of the equations describing the dynamics of an AUV moving into the water are described in [13]. Such equations are highly nonlinear; the non-linearity is additionally accentuated if a model of the propulsion system is taken into account.

The system model suitably developed by the authors constitute a trade-off between kinematic and dynamic AUV model proposed in [13]. In particular, an only kinematic model simplifies too much the real behavior of the vehicle and thus is unsuitable to accurately describe the evolution of its navigation state; on the other hand, a complete dynamic model, including all the physical phenomena acting on a rigid body moving in water, depends on a very high number of parameters, which are often difficult to identify; hence, uncertainty on such parameters would negatively affect the performance of any navigation filter. Then, a mixed kinematic and dynamic model has been adopted considering the dynamical effects only along the longitudinal axis. The equation of motion for the vehicle along the longitudinal degree of freedom has been obtained from [13] by selecting the corresponding elements of the involved matrices. The resulting propagation model for the state variables results as it follows:

$$\begin{bmatrix} [\boldsymbol{\eta}_1]_{k+1} \\ [\boldsymbol{\nu}_1]_{k+1} \end{bmatrix} = \begin{bmatrix} [\boldsymbol{\eta}_1]_{k} \\ [\boldsymbol{\nu}_1]_{k} \end{bmatrix} + \\ + \Delta T \cdot \begin{bmatrix} R\,[\boldsymbol{\nu}_1]_k \\ \begin{bmatrix} \frac{\tau_x([\boldsymbol{\nu}_1]_k, \boldsymbol{u}_k) + F_x([\boldsymbol{\nu}_1]_k)}{m} \\ 0 \\ 0 \end{bmatrix} \end{bmatrix} + \boldsymbol{w}_k \,, \tag{5.1}$$

where $\boldsymbol{\eta}_1$ and $\boldsymbol{\nu}_1$ are, according to [13], respectively the position with respect to a North-East-Down (NED) fixed frame and the linear velocity expressed within a local frame (body frame) fixed with respect to the vehicle; ΔT is the working period of the filter; m is the first row and first column element of the inertia matrix calculated as the sum of the mass of the vehicle itself and the added mass contribution; R is the rotation matrix that represents the orientation of the body frame with respect to the NED one, its estimation is obtained through an independent filter aimed at determining the orientation, see [3] and [5]; τ_x is the force exerted on the vehicle by a set of propellers; $\mathbf{u}_k$ is the set of control inputs given to the systems (i.e. the rotational speeds of the propellers); F_x is the hydrodynamic damping force acting on the longitudinal degree of freedom. Additive white process noise $\boldsymbol{w}_k$ has been taken into account to model uncertainty. Only longitudinal dynamics has been taken into account because AUVs often possess a torpedo shape and travel along the longitudinal direction (the direction of minimal resistance) in order to preserve energy. With the state vector defined in (5.1), and considering that the set of sensors that a typical AUV is equipped with (GPS, Doppler Velocity Log, Depth Sensor, Acoustic Positioning Sensors) provides a direct measurements of its components, the observation function is affine.

Particular attention needs to be given to the terms F_x and τ_x in (5.1): those quantities depend indeed on the particular choice of the propulsion system model and on a-priori unknown coefficients which can be experimentally determined. To obtain an analytical expression for F_x, the starting point is the complete dynamic model of an AUV reported in [13].

Under the assumption of uniform motion (constant speed) and considering a symmetric torpedo-shaped AUV navigating on a horizontal plane, the steady state behaviour of the vehicle model can be modelled through the equilibrium equation:

$$D(\boldsymbol{\nu})\boldsymbol{\nu} = \boldsymbol{\tau}(\boldsymbol{\nu}, \mathbf{u}), \tag{5.2}$$

being D the drag matrix, $\boldsymbol{\nu}$ and $\boldsymbol{\tau}$, respectively, the vector composed of body-fixed linear and angular velocity of the vehicle and the vector of total forces and moments acting on the AUV. Damping force contributions, in the x direction of the body-fixed reference frame are expressed in (5.3) [14]:

$$D_x(v_a) = \frac{1}{2}\rho A v_a C_x = -F_x\ , \tag{5.3}$$

where ρ $[\frac{kg}{m^3}]$ is water density, A $[m^2]$ is the reference area of the AUV, v_a $[\frac{m}{s}]$ is the advance speed along the body-fixed x-axis and C_x is a dimensionless damping coefficient of the vehicle along x (which needs to be identified). According to [15], when dealing with long torpedo-shaped bodies, the reference area of the AUV can be computed as the product of the vehicle length times its diameter $A = Ld$.

For what concerns τ_x, in recent works the authors derived an analytical expression for the model of the propulsion system of an AUV, based on the experimentally validated propeller behavior presented in [16]. According to [17], the thrust T_i generated by a propeller for given advance velocity $v_{a,i}$ (along the motor axis) and rotational speed u_i, is expressed by the following relation:

$$T_i\left(v_{a,i}, u_i\right) = sgn\left(u_i\right)\left(K_0 u_i^2 - \frac{K_0\left|u_i\right| g\left(sgn\left(u_i\right) v_{a,i}\right)}{p}\right) \tag{5.4}$$

where $g\left(\cdot\right)$ is a piecewise function defined as

$$g\left(x\right) = \begin{cases} 0 & for\ x \leq 0 \\ x & for\ 0 < x \leq \left|u_i\right| p \\ \left|u_i\right| p & for\ x > \left|u_i\right| p \end{cases} \tag{5.5}$$

being p the propeller pitch, and K_0 the thrust coefficient which relates motor thrust and propeller speed when the advance speed v_a is null (bollard thrust. In [18], it has been shown that the relationship between $T_i(0, u_i)$ is approximately parabolic:

$$K_0 = \left.\frac{T_i}{u_i^2}\right|_{v_{a,i}=0}. \tag{5.6}$$

Once the expression of all T_i, $i = 1, ..., m$, where m is the number of the propllers of the vehicle, is known, the vector of forces and moments acting on the AUV can be computed. Let

$$\mathbf{U} = [T_1\ T_2\ \cdots T_i \cdots T_m]^T \in \mathbb{R}^m, \quad T_i = T_i(v_{a,i}, u_i) \tag{5.7}$$

denote the vector collecting the thrusts exerted by the m propellers of the vehicle; then, the vector of force and moment acting on the AUV $\boldsymbol{\tau}(\boldsymbol{\nu}, \mathbf{u})$ is computed following the linear model:

$$\boldsymbol{\tau}(\boldsymbol{\nu}, \mathbf{u}) = B\mathbf{U}(\boldsymbol{\nu}, \mathbf{u})\,. \tag{5.8}$$

Matrix B is constant and dependent on the vehicle geometry, and it is expressed as follows:

$$B = \begin{bmatrix} A_1 \\ A_2 \end{bmatrix}, \tag{5.9}$$

with

$$A_1 = \left[\cdots \mathbf{n}^b_{mi} \cdots\right], \quad A_2 = \left[\cdots \left(\mathbf{P}^b_{mi} - \mathbf{O}^b\right) \times \mathbf{n}^b_{mi} \cdots\right], \tag{5.10}$$

where $\mathbf{P}^b_{mi}$ is the position of the i-th motor, expressed in the body frame with origin $\mathbf{O}^b$, and $\mathbf{n}^b_{mi}$, is the line of action of the i-th motor expressed in the body frame.

By performing suitable bollard thrust measurements at different propeller rotational speeds, the value K_0 for a given motor and propeller pair can be determined. Once K_0 is known, its value can be used for the determination of the longitudinal drag coefficient C_x. Let us consider, for example, a torpedo-shaped AUV with two rear propellers (whose lines of action can be assumed parallel and coincident with the body-fixed x-axis) moving in the forward direction at the velocity v_a. In addition, let u indicate the (same) rotating speed of such propellers. Then, by using (5.4)–(5.5), (5.2) can be rewritten as follows:

$$2\left(K_0 u^2 - \frac{K_0 n v_a}{p}\right) = \frac{1}{2}\rho L d {v_a}^2 C_x\,, \tag{5.11}$$

which can then be solved for C_x:

$$C_x = 2\,(2K_0)\,\frac{(u^2 - u\frac{v_a}{p})}{\rho L d {v_a}^2}\,. \tag{5.12}$$

If several longitudinal motion tests are performed, with the AUV moving at different speeds v_a and for measurable u, a least square estimate of C_x can then be determined.

5.2.2 *Marine Current Estimation*

Sea currents can considerably affect the motion of an AUV. The research idea here proposed is using onboard sensors and the knowledge of the vehicle dynamics to get an online estimation of the local marine currents. With respect to (5.1), the state vector dimension has been augmented (from six to eight elements) to include the two horizontal components of the sea currents expressed in the fixed frame (u_c^E North component and v_c^E East component); then, the model has been properly modified in order to consider the effect of the currents on vehicle dynamics. In this context, the vertical component of the current has not been considered (its effect on the vehicle motion can be reliably observed through depth sensor measurements). For sea currents, a locally constant model has been considered with an additive white process noise included in the vector $\boldsymbol{w}_k$ to take into account model uncertainties (slowly varying current velocity). The vehicle dynamics can be expressed as a function of the relative velocity $\boldsymbol{\nu}_{1r} := \boldsymbol{\nu}_1 - \boldsymbol{\nu}_{1c}$, where $\boldsymbol{\nu}_{1c}$ is the current velocity expressed in the body frame:

$$\boldsymbol{\nu}_{1c} = R^T \begin{bmatrix} u_c^E \\ v_c^E \\ 0 \end{bmatrix} \tag{5.13}$$

According to the above-mentioned assumptions, the proposed model is:

$$\begin{bmatrix} [\boldsymbol{\eta}_1]_{k+1} \\ [\boldsymbol{\nu}_1]_{k+1} \\ \begin{bmatrix} u_c^E \\ v_c^E \end{bmatrix}_{k+1} \end{bmatrix} = \begin{bmatrix} [\boldsymbol{\eta}_1]_k \\ [\boldsymbol{\nu}_1]_k \\ \begin{bmatrix} u_c^E \\ v_c^E \end{bmatrix}_k \end{bmatrix} + \tag{5.14}$$

$$+ \Delta T \cdot \begin{bmatrix} R\,[\boldsymbol{\nu}_1]_k \\ \begin{bmatrix} \frac{\tau_x([\boldsymbol{\nu}_{1r}]_k, \boldsymbol{u}_k) + F_x([\boldsymbol{\nu}_{1r}]_k)}{m} \\ \frac{F_y([\boldsymbol{\nu}_{1r}]_k)}{m} \\ 0 \end{bmatrix} \\ \begin{bmatrix} 0 \\ 0 \end{bmatrix} \end{bmatrix} + \boldsymbol{w}_k$$

where F_y is the hydrodynamic damping force acting on the lateral degree of freedom.

At the moment, the UKF algorithm has been validated offline exploiting the signals collected by Typhoon AUV [12], an AUV developed and built by the University of Florence. The UKF is based on the Unscented Transform (UT), a derivative-free technique capable of providing a more accurate statistical characterization of a Random Variable (RV) undergoing a nonlinear transformation [6]. In particular, the UT is a deterministic technique suited to provide mean and covariance of a given RV subjected to a nonlinear function given a minimal set of its samples. The adopted implementation is the one proposed in [19]. Its onboard sensors are a Doppler Velocity Log—DVL, an Attitude and Heading Reference System—AHRS, a Depth Sensor and a Global Positioning System—GPS on the surface; the data used in this paper were logged on the archaeological site of Cala Minnola, Levanzo Island in Sicily (Italy) during Summer 2015 (during the first final demonstration of the European ARROWS project [20]). All the sensor noise contributions are modelled as zero-mean Gaussian signals. Because of sensors different working frequencies and their possible unavailablity (for instance, the GPS), the number and the type of measurements to be used in the filter correction step may be different. This is taken

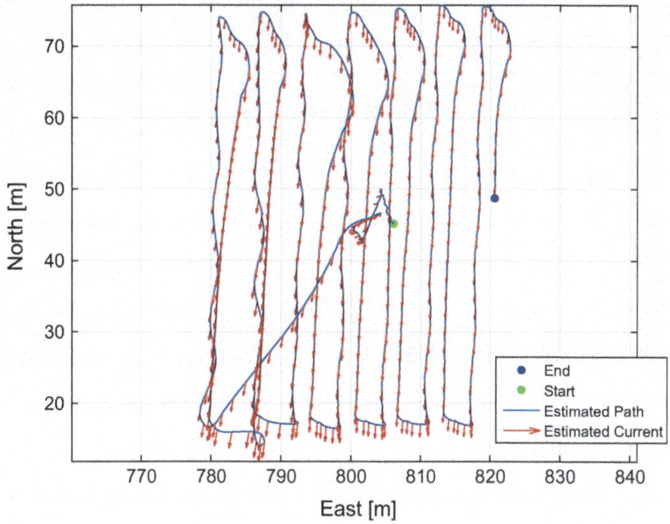

Fig. 5.1 Simultaneously estimated path and sea current for the considered mission, start point coordinates (N37.99623 E12.34615)

into account defining a coherent measurement model at each iteration according to the updated sensor signals.

A non-negligible current from North was present on surface during the day of the mission. According to this, in order to save energy onboard taking into account the torpedo-shape of Typhoon AUV, the legs of the planned path were oriented almost parallel to the North-South direction. The mission lasted 2168 s with a starting point indicatively in the center of the target area (green dot in Fig. 5.1). Typhoon AUV reached the desired altitude (2 m from the sea bottom, corresponding to a depth of about 25 m), then it navigated along the path controlling its altitude and finally it surfaced on the ending point (blue dot). In Fig. 5.1 the estimated path is shown along with the estimated direction of the sea current represented by red vectors at intervals of 5 s (for the sake of plot intelligibility). In Figs. 5.2 and 5.3 the estimated direction and intensity of the sea current are visible. A stable trend starts approximately from the third minute of the mission. The mean value of the estimated bearing is about 185°, with respect to the North, whereas the value of the speed magnitude is between 0.15 and 0.20 m/s. After the first transient, corresponding to the immersion phase, the estimated direction of the sea current remains approximately at a constant value coherent with what expected from the on-field experience.

From the results analysis, both the direction and the intensity of the estimated current are coherent with what observed during the experimentation, but a deeper validation of the proposed algorithm is scheduled for the near future: independent current measurements will be used first to tune the algorithm and then as ground truth to have a quantitative evaluation of the estimation performance.

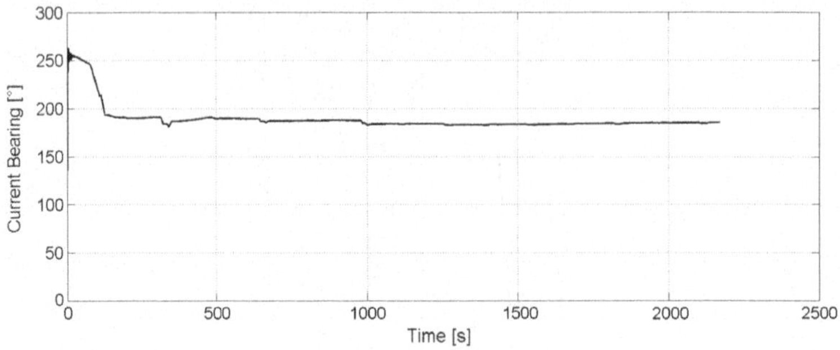

Fig. 5.2 Estimated direction of the sea current

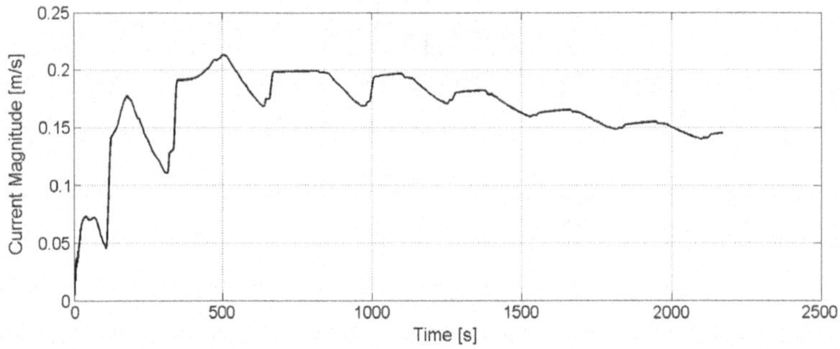

Fig. 5.3 Estimated intensity of the sea current

5.3 Acoustic Positioning Systems for Navigation

Past works of the authors focused on the investigation of different acoustic sensor networks for localisation purposes aimed at improving the navigation performance by providing the AUV with a measurement of relative position. Among the different developed and tested configurations, two of them are summarized in this work, in the following subsections. The research outcome here proposed, if combined with the results of the previous sections, is the base to achieve an autonomous system characterized with long term navigation capabilities without limits on its operation due to the necessity of periodic surface procedures to get GPS. Moreover, it can provide an improvement of the quality of the collected data in terms of their geo-referencing.

5.3.1 Cooperative Navigation

The configuration described in this subsection is thought to be adopted for scenarios including more AUVs cooperating for a common mission on the same operational

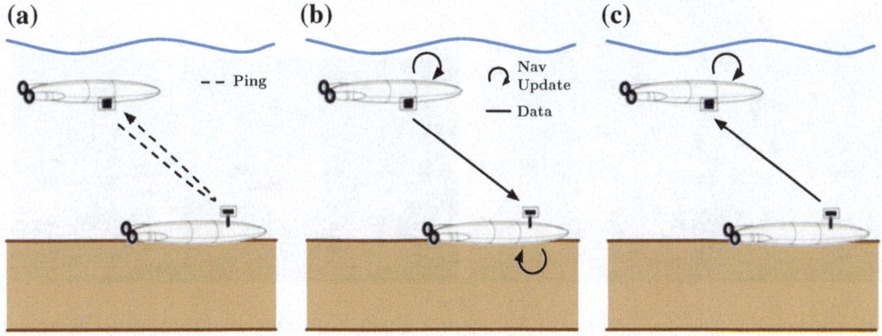

Fig. 5.4 The cooperative step of the navigation algorithm between two vehicles. **a** Localisation: the team master pings a slave vehicle to get the USBL measurement. **b** Master update, data transmission and slave update: the team master exploits the USBL measurement to localise itself and the slave, updating the overall navigation state. After the correction, the team master sends the refined position to the slave, which employs the received information to correct its navigation status. **c** Data transmission and final update: when its time slot arrives, the slave vehicle sends back its current navigation status; finally, the team master uses the received estimate to perform a further correction of the navigation status of the slave

area. For this kind of application it is common that the vehicles are equipped with acoustic modems for the purpose of communication among themselves. The idea is to equip one of the AUVs with an USBL system capable of localizing acoustic modems. The vehicle periodically surfaces to get a reliable measurement of its absolute position and obtains, through integration with USBL measurements, information about the absolute position of the other AUVs. By means of communication, the position measurements are then transmitted to the corresponding vehicles that can use it to improve the navigation estimation. Figure 5.4 describes the scenario.

A first test to be considered a proof of concept for the proposed approach took place in 2014 during Breaking the Surface 2014 workshop in Biograd na Moru, Croatia in the framework of a demonstration of the FP7 European ARROWS project [20]. The test involved two Typhoon class AUVs developed during the Tuscany Region THESAURUS project [21], namely TifOne—slave vehicle, equipped with a localizable acoustic modem and navigating underwater controlling its path along a preplanned set of waypoints—and TifTwo—master vehicle, equipped with an USBL sensor and holding a constant position on the surface. All the measurements were logged and fused off-line in an Extended Kalman Filter (EKF) based on a kinematic model to verify the developed approach—the reader may refer to [11] for the implementation details. Figure 5.5 describes the target path with a null reference depth for the first two lines. Figure 5.6 reports the output of the navigation filter using the acoustic positioning measurements as corrections.

In Table 5.1, the performance of the cooperative algorithm in terms of the final localisation error of the TifOne AUV with respect to the GPS position at the surfacing (indicated in Fig. 5.5 with the magenta circle) are compared with the non-cooperative case for several vehicle configurations in the three performed trials. In particular, in

Fig. 5.5 Left reference path defined by 5 waypoints. Right the two involved vehicles—TifOne and TifTwo—moored before the beginning of the mission

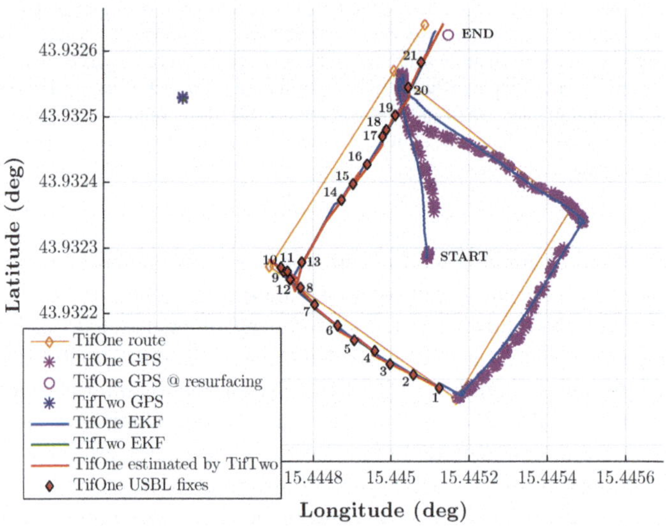

Fig. 5.6 Estimated path based on acoustic position measurements as corrections

the first configuration (first two rows) TifOne is supposed to mount the DVL perfectly aligned with the body-fixed frame and to employ the measured velocity in the navigation filter. In this case the performance of both the algorithms are comparable, with a slightly bettering in the cooperative scheme: the position error is always kept close to the GPS precision in all the three runs. The only exception is represented by the third mission, in which the error is higher with the introduction of cooperation but the number of USBL fixes is smaller than in the other runs (21 fixes for the first experiment, 22 for the second one and 11 for the third one). Moreover, in the last part of the third trial no USBL corrections were received: the TifOne estimation filter cannot take advantage of the cooperative part. The second configuration (third and fourth rows) considers TifOne still using DVL corrections, but simulating a misalignment of 1° with respect to the body frame along the z axis. In this case the benefits

Table 5.1 Cooperative navigation: TifOne error at the final surfacing with several configurations

TifOne configuration			Error at the surfacing (m)		
Cooperation	DVL	DVL-body misalignment (°)	1st run	2nd run	3rd run
X	✓	0.0	4.6	3.7	3.8
✓	✓	0.0	2.7	1.1	4.2
X	✓	1.0	5.7	4.6	4.8
✓	✓	1.0	2.8	1.2	4.5
X	X	–	141.2	140.9	254.3
✓	X	–	3.7	2.3	20.7

of the proposed cooperative navigation strategy over the selfish one are evident: the former exhibits performance similar to the previous, ideal configuration, whereas the latter shows an increase of the 25% in the final localisation error. Finally, in the third configuration here considered (last two rows) TifOne is supposed not to have the DVL on-board, i.e. the velocity corrections are not used in the estimation filter. This is the case in which the slave vehicle self-localisation relies only on dead-reckoning. While the position estimation error based on proprioceptive sensors grows indefinitely, as expected, the cooperative algorithm is able even in this case to bound the horizontal position drift and to keep the localisation error within the GPS precision.

The experiment demonstrates that it is possible to adopt the proposed configuration to maintain the estimation error limited by feeding the AUVs with position measurements, even with a model that does not take into account the dynamics of the vehicle. The two main advantages can be summarized as:

- **no dependence from the support vessel**—the proposed architecture for the acoustic sensor network ensures the result independently from the presence of a support vessel. This permits the AUV team to be really autonomous also in case of lost of connection with command and control station.
- **no necessity of device deployment**—the considered scenario is completely independent from fixed acoustic nodes as it would be necessary in a LBL system with a double advantage. It permits not to dedicate time and economic resources to deployment and calibration procedures. In addition it does not impose for the AUV team a spatial constraint on the operational area.

5.3.2 *Acoustic SLAM*

The theoretical basics of an acoustic based Simultaneous Localisation And Mapping (SLAM) strategy to maintain bounded the position estimation error were recently developed [12]. The proposed approach is applicable even for a single vehicle and it is based on the presence of one or more acoustic nodes underwater in a static position. The vehicle houses on board and USBL sensor mounted in the inverse configuration (iUSBL) and during its mission gets relative position measurements of the different nodes. The state vector of the navigation filter contains the coordinates of the acoustic nodes that are estimated in real-time. Through this implementation, the static acoustic nodes play the role of the landmarks of a classic SLAM framework. A map of the landmarks is estimated in real-time as well as the position of the vehicle with the respect to it. The advantage of the proposed technique with respect to the other ones based on the deployment of acoustic devices is represented by the complete independence from any initial calibration procedure. As soon as the devices are deployed, the vehicle can start navigating estimating their position during the mission with considerable advantages in terms of costs and time. The possible applications of the proposed method range from a completely unknown environment with the acoustic devices deployed in generic positions to a structured environment where the nodes can be resident devices. An off-line validation based on experimental data were carried out. The occasion to collect appropriate data for the purpose was the CommsNet13 cruise, organized and scientifically coordinated by the NATO S&T Ctr. for Maritime Research and Experimentation (CMRE, formerly NURC). The experiment took place in September 2013 in the La Spezia Gulf, North Tyrrhenian Sea. In that occasion, the Typhoon vehicle (TifTwo) was equipped with the USBL modem, a GPS antenna and a low cost, low accuracy Inertial Measurement Unit (IMU). The USBL modem could communicate with an ad-hoc installation of fixed sensors consisting of some battery-operated acoustic modems. Two different approaches—different for the a priori knowledge of the number of the nodes of the underwater acoustic network—were investigated. In the case that this information is known, an extended state vector including the coordinates of the acoustic nodes is considered. Whereas, in the case that it is not known, a dynamic resizing of the state vector based on newly "met" nodes takes place. Implementation details can be found in [12]. Results from both the configuration are summarized in the plots of Figs. 5.7 and 5.8. The considered ground truth is the measured GPS deploying position for the 6 nodes of this experiment (crosses in the plots) and the GPS signal for the AUV (magenta asterisks) that was on the surface during all the mission. GPS measurements are not used within the navigation filter during all the mission, but just at the beginning for initialization purposes. Figures 5.9 and 5.10 report the estimation error with respect to the ground truth both for the vehicle and for the acoustic nodes. For both the cases, the estimation error remains bounded and within 3–4 times GPS accuracy. The result is even more important if considering that it is obtained with a low cost inertial motion units and in presence of hostile acoustic channel conditions (with the vehicle, and thus the USBL sensor, close to the surface the acoustic localisation system performance are affected by strong reflection and multi-path phenomena).

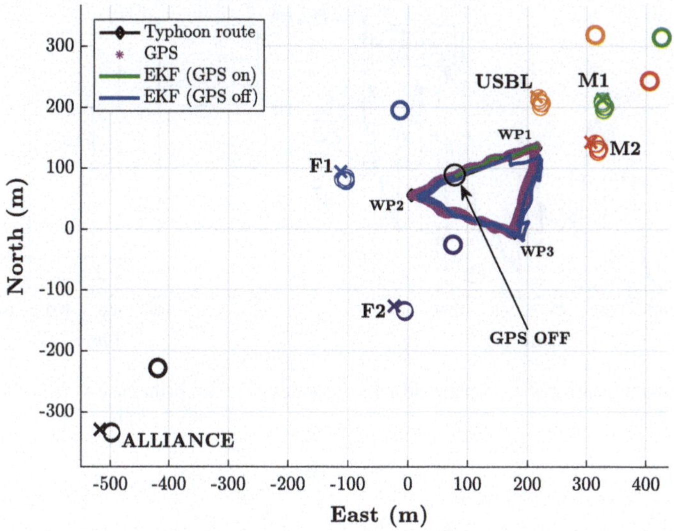

Fig. 5.7 Estimated and ground truth AUV path and acoustic nodes positions in the case of a priori knowledge of their number

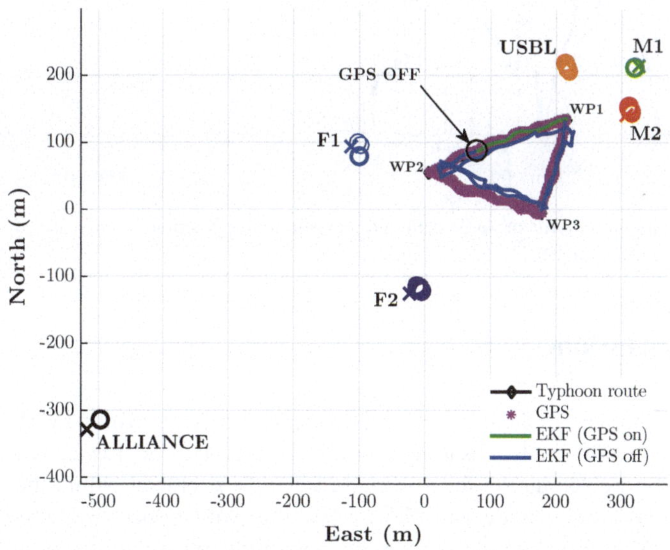

Fig. 5.8 Estimated and ground truth AUV path and acoustic nodes positions in the case of no a priori knowledge of their number

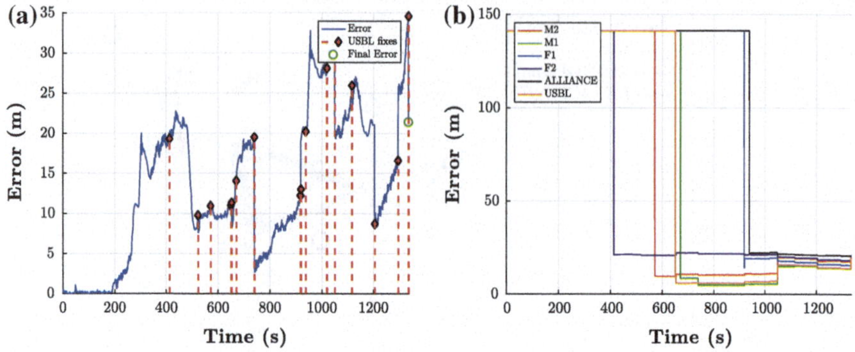

Fig. 5.9 Estimation error of AUV path and acoustic nodes positions in the case of a priori knowledge of their number

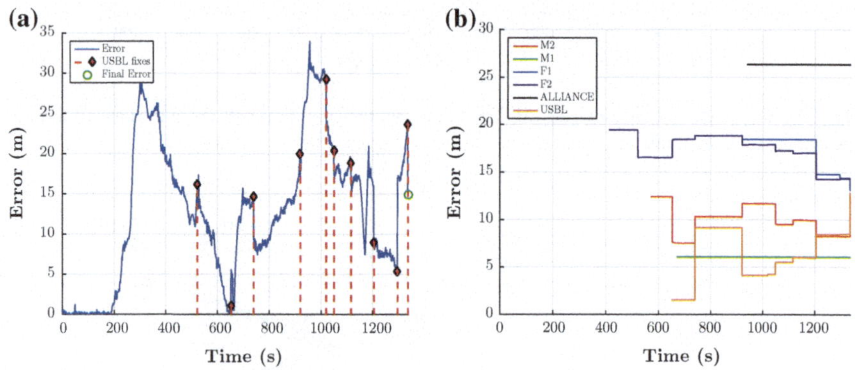

Fig. 5.10 Estimation error of AUV path and acoustic nodes positions in the case of no a priori knowledge of their number

5.4 Conclusions

This work summarizes the research developments carried out within ISME about the specific navigation and localisation topics for Unmanned Underwater Vehicles. The problem of pose estimation for this kind of vehicle is crucial within the complete motion control architecture and plays a key role with respect to the capacity of payload data geo-referencing. Without the possibility of exploiting a local acoustic-based positioning system, despite the quality of involved sensors, the estimation error is characterized by an unbounded drift. The strategy described in this work is twofold: part of the research is oriented to the reduction of the growth velocity for the estimation error and the other part is dedicated to the investigation of different acoustic-based sensor networks for the error reset.

As it concerns the increasing rate of the estimation error, the described methodologies to reduce it are based on (a) the design of the dynamic model of the system including the main involved phenomena (e.g. sea current estimated in real-time) for the prediction of its behaviour (b) the investigation of the adoption of the Unscented Kalman Filter as an alternative to the more common approach based Extended Kalman Filter for the estimation a process whose dynamics is intrinsically nonlinear.

The two strategies described in this work to provide the vehicle with a position measurement are thought for different applications: (a) a solution based on an integrated communication and localisation system oriented to make a team of cooperating vehicles independent from the support vessel without the necessity of dedicated devices deployment (b) a framework inspired to Simultaneous Localisation And Mapping applications for ground robotics has been theorized and validated exploiting experimental data; this approach is based on the deployment of static acoustic nodes (SLAM landmarks) that do not require a calibration procedure as their positions are mapped during the mission execution.

Future steps are oriented in fusing the different approaches presented in this work, till the moment independently experimented for validation purposes, for the implementation on ISME vehicles of a complete navigation strategy potentially valid for applications that require persistent autonomy as e.g. hydrography or geophysics oriented missions.

Acknowledgements The described work has been partially supported by ARROWS project (it has received funding from the European Union Seventh Framework Programme for research, technological development and demonstration under grant agreement no. 308724), BRUCE project (it has received funding from the European Union's Seventh Framework Programme within "SUNRISE—Open Call 2 for new beneficiaries", under grant agreement no. 611449), WiMUST project (it has received funding from the European Union's Horizon 2020 research and innovation programme under the grant agreement no. 645141), THESAURUS project (PAR FAS REGIONE TOSCANA Linea di Azione 1.1.a.3).

References

1. B. Allotta, A. Caiti, R. Costanzi, F. Fanelli, D. Fenucci, E. Meli, A. Ridolfi, A new AUV navigation system exploiting unscented Kalman filter. J. Ocean Eng. **113**, 121–132 (2016)
2. L. Paull, S. Saeedi, M. Seto, H. Li, AUV navigation and localization: a review. IEEE J. Ocean. Eng. **39**(1), 131–149 (2014)
3. R.E. Mahony, T. Hamel, J.M. Pflimlin, Nonlinear complementary filters on the special orthogonal group. IEEE Trans. Autom. Control **53**(5), 1203–1218 (2008)
4. B. Allotta, R. Costanzi, F. Fanelli, N. Monni, A. Ridolfi, Single axis FOG aided attitude estimation algorithm for mobile robots. J. Mechatron. **30**, 158–173 (2015)
5. R. Costanzi, F. Fanelli, N. Monni, A. Ridolfi, B. Allotta, An attitude estimation algorithm for mobile robots under unknown magnetic disturbances. IEEE/ASME Trans. Mechatron. **21**, 1900–1911 (2016)
6. S. Julier, J. Uhlmann, Unscented filtering and nonlinear estimation. Proc. IEEE **92**, 401–422 (2004)
7. B. Allotta, A. Caiti, L. Chisci, R. Costanzi, F. Di Corato, F. Fanelli, C. Fantacci, D. Fenucci, E. Meli, A. Ridolfi, A comparison between EKF-based and UKF-based navigation algorithms for AUVs, in *Proceedings of the OCEANS 2015* (MTS/IEEE, Genova, 2015). https://doi.org/10.1109/OCEANS-Genova.2015.7271681

8. Y. Bar-Shalom, X.R. Li, T. Kirubarajan, *Estimation with Applications to Tracking and Navigation: Theory Algorithms and Software* (Wiley, New York, 2001)
9. A. Bahr, J.J. Leonard, M.F. Fallon, Cooperative localization for autonomous underwater vehicles. Int. J. Robot. Res. **28**(6), 714–728 (2009)
10. T.C. Furfaro, J. Alves, An application of distributed long baseline node ranging in an underwater network, in *Proceedings of UComms 2014, Sestri Levante, Italy* (2014)
11. B. Allotta, A. Caiti, R. Costanzi, F. Di Corato, D. Fenucci, N. Monni, L. Pugi, A. Ridolfi, Cooperative navigation of AUVs via acoustic communication networking: field experience with the Typhoon vehicles. Auton. Robot **40**, 1229 (2016)
12. A. Caiti, F. Di Corato, D. Fenucci, B. Allotta, R. Costanzi, N. Monni, L. Pugi, A. Ridolfi, Toward underwater acoustic-based simultaneous localization and mapping. Experimental results with the Typhoon AUV At Commsnet13 Sea Trial, in *Proceedings of the OCEANS'14* (MTS/IEEE St. Johns, 2014)
13. T.I. Fossen, *Guidance and Control of Ocean Vehicles* (Wiley, New York, 1994)
14. E.D. Barros, A. Pascoal, E. de Sac, Investigation of a method for predicting AUV derivatives. J. Ocean Eng. **35**(16), 1627–1636 (2008)
15. B.R. Munson, T.H. Okiishi, W.W. Huebsch, A.P. Rothmayer, *Fundamentals of Fluid Mechanics* (Wiley, New York, 2013)
16. J. Carlton, *Marine Propellers and Propulsion* (Elsevier, Oxford, 2007)
17. B. Allotta, A. Caiti, L. Chisci, R. Costanzi, F. Di Corato, C. Fantacci, D. Fenucci, E. Meli, RIdolfi A, An unscented Kalman filter based navigation algorithm for autonomous underwater vehicles. J. Mechatron. **39**, 185–195 (2016)
18. B. Allotta, F. Bartolini, R. Costanzi, N. Monni, L. Pugi, A. Ridolfi, Preliminary design and fast prototyping of an autonomous underwater vehicle propulsion system. Proc. Inst. Mech. Eng. (2014). https://doi.org/10.1177/1475090213514040
19. E.A. Wan, R. Merwe, *The Unscented Kalman Filter* (S. Haykin Edition, New York, 2001)
20. B. Allotta, R. Costanzi, A. Ridolfi et al., The ARROWS project: adapting and developing robotics technologies for underwater archaeology, in *Proceedings of the IFAC Workshop on Navigation, Guidance and Control of Underwater Vehicles (NGCUV2015) Girona* (2015)
21. A. Caiti, V. Calabrò, F. Di Corato, T. Fabbri, D. Fenucci, A. Munafò, B. Allotta, F. Bartolini, R. Costanzi, J. Gelli, N. Monni, M. Natalini, L. Pugi, A. Ridolfi, Thesaurus: AUV teams for archaeological search. Field results on acoustic communication and localization with the Typhoon, in *22nd Mediterranean Conference on Control and Automation Palermo* (2014), pp. 857–863

Chapter 6
Design and Control of an Autonomous Underwater Vehicle (AUV-UMI)

Adrian Manzanilla, Miguel Garcia, Rogelio Lozano and Sergio Salazar

Abstract This paper outlines the conception, modeling and control of a rover type modular mini submarine. The development and implementation of the mechanical structure as well as the embedded electronics is described. The onboard instrumentation and sensors required to collect data on the environment and on its own position and orientation are also described. The mathematical representation to describe the movement of an underwater vehicle is analyzed considering the characteristics and limitations of the underwater robot. Furthermore, a control algorithm is implemented based on Lyapunov theory and Backstepping Integral Adaptive (BIA). This control strategy stabilizes the vehicle in position and orientation. The proposed control algorithm is validated in numerical simulations as well as in experimental tests which confirm the good performance of the prototype and the controller.

6.1 Introduction

The technological advances in the last years have made possible the development of a diversity of autonomous underwater vehicle (AUV) as important tools in the oceanographic investigation and different civil and military applications. In the literature, several works can be found that present the design and development of submarine robots, as shown in [2, 6, 22] which present the construction of low

A. Manzanilla · M. Garcia · S. Salazar
UMI LAFMIA CNRS 3175, CINVESTAV-IPN, Ave. IPN 2508,
07360 Mexico City, Mexico
e-mail: amanzanilla@cinvestav.mx

M. Garcia
e-mail: miguelgarciar@cinvestav.mx

S. Salazar
e-mail: sergio.salazar.cruz@gmail.com

R. Lozano (✉)
Sorbonne Universités, Université de Technologie de Compiègne,
CNRS, UMR 7253, Compiègne, France
e-mail: rlozano@hds.utc.fr

L. Jaulin et al. (eds.), *Marine Robotics and Applications*, Ocean Engineering
& Oceanography 10, https://doi.org/10.1007/978-3-319-70724-2_6

cost autonomous mini submarines to be used as test benches. Other authors propose the design of torpedo-type submarine vehicles for applications that require greater autonomy as in [1]. More complex bio-inspired systems using biomimetic flap are presented in [11] or equipped with vectorized thrusters that allow the vehicle to orient both the propeller and the rear fins in three degrees of freedom are presented in [3]. Nowadays, the design of autonomous modular vehicles with the capability to perform different tasks is proposed in [15] which presents the development of an underwater robot with the ability to reconfigure the number of propellers and their location to generate motions having up to six degrees of freedom. Several models are proposed in the literature to analyze the translational and rotational movements of the submarine. An example is the model proposed by Graver for glider type vehicles [7] and the matrix model proposed by Fossen where the properties of the matrices are used [19].

The ocean is an environment in which operating an underwater vehicle faces some difficulties. The control strategy refers to the methodology used to manipulate an underwater vehicle, stabilize it for performing a mission, even in the presence of perturbations and uncertainties in the model. In the literature, several works can be found that present different control techniques to stabilize an underwater robot in both position and orientation. In [20] the design of a PD controller with saturations to stabilize an autonomous underwater vehicle that moves at low speed is proposed. On the other hand [12] compares the performance of a PD controller and an adaptive nonlinear state feedback control to regulate one degree of freedom in an underwater vehicle. It was shown that adaptive effect gives greater robustness by compensating external disturbances. However, [4, 5, 9] present the design of controllers based on Lyapunov theory and backstepping techniques to stabilize AUVs. In recent years, controllers with adaptive effects have been developed to give robustness with respect to model uncertainties and external shocks, examples of this are shown in [10, 14]. Other proposals include the use of sliding-mode techniques for controller design, taking into account the unknown perturbations, nonlinearities and uncertainties in the model, as in [8, 16, 21].

The goal of this article is the design of a low cost submarine robot that is adapted to perform different applications. The analysis of the vehicle movement taking into account its characteristics and limitations is presented. Furthermore a control algorithm is proposed based on the theory of Lyapunov and backstepping techniques which stabilizes the vehicle even in the presence of unknown external disturbances.

6.2 Prototype Design

It is possible to distinguish several stages in the development of the prototype. The first stage focuses on the design of the AUV using materials that are easy to replace in such a way that the vehicle is modular. The objective is to conceptualize the design through the use of Computer Aided Design software (SolidWorks), with a structure based on a nylamid and polyethylene with three acrylic tubes of 6 in.

Fig. 6.1 CAD design, AUV-UMI vehicle

diameter that are pressure and corrosion resistant, among other physical properties. Since an underwater robot must provide a compartment to contain its electronic components in a dry and watertight environment, it is therefore proposed to use two aluminum caps with O-rings, fastened by means of four stainless steel screws. In addition we are using SEACON dry connectors, as shown in Fig. 6.1.

The prototype allows easy assembly of external devices and components, see Fig. 6.2. Its main characteristics can be summarized as follows: The vehicle dimensions are 63 × 63 × 26 cm with a weight in air of 33.9 kg, time of autonomy of approximately six hours and a depth operation capability of 32 m with option to 100 m. These dimensions are for transportation logistics, and the diameter of the tubes are designed to be able to introduce the embedded system. However, it is important to highlight some considerations in the conception of this prototype, by design the roll angle is mechanically stable, the origin of the body structure O_B coincides with the center of gravity C_G and the submarine robot has two axes of symmetry.

6.2.1 Embedded System

The electronic architecture of the AUV-UMI is a relatively simple architecture based on an INTEL NUC computer, which computes the proposed algorithms and acquires sensor data for the vehicle to operate in ROV or AUV mode. The perception of the robot environment is done using several sensors as an inertial measurement unit (9DOF Razor IMU) by which the orientation and angular velocity of our vehicle is known. The pressure sensor allows to calculate the depth, two cameras (PlayStation Eye) are used to acquire information of the images, however, its use depends on what

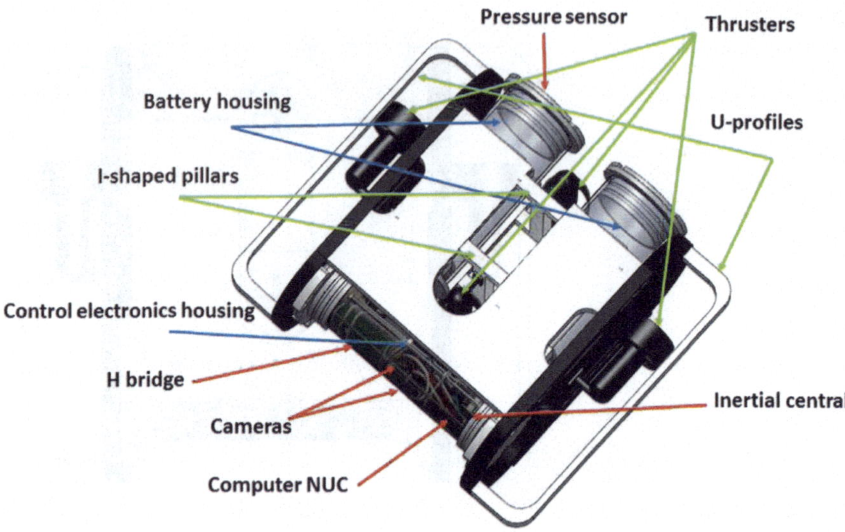

Fig. 6.2 Internal structure AUV-UMI

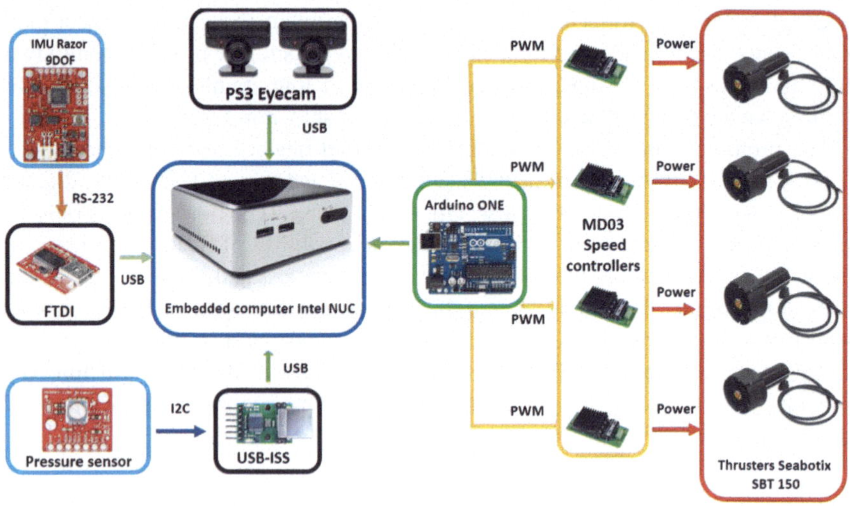

Fig. 6.3 Embedded system AUV-UMI

kind of mission is assigned to the vehicle. The arduino-one is used as a communication interface between the speed controllers and the built-in computer, while the USB-ISS communicates all devices using the I2C communication protocol to the computer, see Fig. 6.3.

6.2.2 Propulsion System

The AUV-UMI is equipped with four thrusters that allow the underwater vehicle to move where the autonomous controller or operator needs to go. Seabotix's BTD150 DC thrusters are electric motors that are coupled to a propeller that produces a thrust effect by moving the fluid from front to rear due to the pressure difference produced. Because of this, the location of the motors was chosen to achieve an active control over four degrees of freedom and passively maintain a zero rotation angle so that the buoyancy center is above the center of mass. In order to generate the appropriate voltage by means of pulse width modulation, a DC motor controller MD03 is used to generate the 4 appropriate signals to manipulate both the magnitude of the speed and the control of the direction generated by the propellers.

6.2.3 Shore-Based Station

In Fig. 6.4, the ground station that is composed of a notebook computer, a joystick and a router can be seen. Using the remote desktop connection application it can manipulate the embedded computer from an external computer through communication via Ethernet and wireless via SSH. The software facilitates to test the operation of the different devices and interfaces with the computer on board the vehicle, and allows changes in the algorithms of control and vision or even configure the mission while the vehicle is on the surface or connected in water ROV mode.

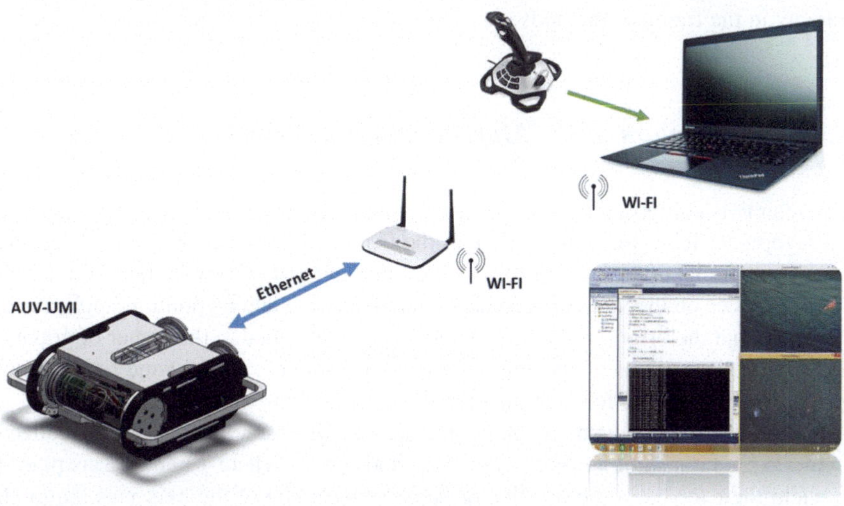

Fig. 6.4 Shore-based station

6.3 Dynamics of Underwater Vehicle

The mathematical model of the system allows to know the equations that describe its behavior to understand and manipulate its dynamics. The representation proposed by Fossen [19] is adopted in this paper to describe the motion of the underwater vehicle, can be represented from the following Newton-Euler motion equations in terms of the body frame to standardize the notation described by the SNAME [17].

$$\begin{aligned} \dot{\eta} &= J(\eta)\nu \\ M\dot{\nu} + C(\nu)\nu + D(\nu)\nu + g(\eta) &= \tau + w \end{aligned} \tag{6.1}$$

where η represents the position and orientation of the vehicle, ν denotes the linear and angular velocity in the frame of the body, $J(\eta)$ is the kinematic transformation between the body and the inertial frame, M represents the inertia matrix (consisting of M_{RB} which represents the mass and inertia matrix of the rigid body, M_A which defines the matrix due to the kinetic energy of the fluid), $C(\nu)$ describes the terms of Coriolis (is composed of the Coriolis matrix of aggregate mass $C_{A(\nu)}$ and by the Coriolis matrix of rigid body $C_{RB}(\nu)$) and restoring forces, $D(\nu)$ describes the hydrodynamic damping matrix (is mainly caused by the damping potential $DP(\nu)$, body friction $DS(\nu)$, wave drift damping $DW(\nu)$ and damping due to the vortex $DM(\nu)$), $g(\eta)$ is the vector of gravitational forces and hydrostatic thrust, τ represents the vector of control inputs and w defines the perturbation vector. In addition, we can denote, $\eta = [x, y, z, \phi, \theta, \psi]^T$, $\nu = [u, v, w, p, q, r]^T$ and $\tau = [X, Y, Z, K, M, N]^T$, where $x, y, z, \phi, \theta, \psi$ define the position and orientation of vehicle in the inertial frame and u, v, w, p, q, r represent the linear and angular velocity in the frame of the body.

6.3.1 *Description of the Movement of the Vehicle*

The vehicle is composed by four thrusters producing forces and pairs, located: two at the top parallel to the axis z_B (one at the front and the other at the rear) and the other two parallel to the axis x_B (one of the side right and one on the left). The effect of these forces on the vehicle depends on its magnitude and its point of application, assuming that the origin of the body structure is established in the center of gravity. As a consequence, it has only four degrees of freedom (x, z, θ, ψ) and four control inputs, where fi is the force of the propeller $i = 1:4$, as shown in Fig. 6.5a.

The translational movement taking into account its characteristics can be realized on two axes of translation in the frame of the body (x, z), where $f1$ and $f2$ are placed to wield their forces in the positive or negative direction of the axis x as required, while $f3$ and $f4$ wield their forces only on the z axis. On the other hand, the rotation movement on the axis of the frame of the body y known as pitch angle, is given by the difference of forces generated between $f3$ and $f4$; where the resulting force

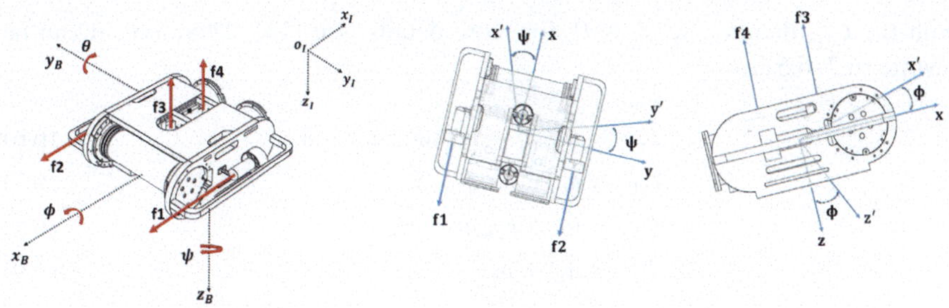

(a) Forces acting in AUV-UMI (b) Rotation and translation movement

Fig. 6.5 Forces and movements in the prototype

is located at the middle of the two motors, in other words exactly perpendicular to the axis x, so it is possible to generate a rotational motion on the axis y. In order to generate a rotation motion on the axis z, also known as yaw angle, it is necessary that one of the forces produced by the side motors is larger than the other, the direction of rotation will depend on which of the two actuators has greater force see Fig. 6.5b.

Developing (6.1), the translational and rotational movements of our robot can be expressed as:

$$
\begin{aligned}
X &= (m - X_{\dot{u}})\dot{u} + X_{\dot{v}}\dot{v} - (my_g - X_{\dot{r}})\dot{r} - (Z_{\dot{w}}w + mw)q \\
&\quad +(Y_{\dot{v}}v + mw)r - (X_u)u + f_B \sin\theta
\end{aligned} \tag{6.2}
$$

$$
\begin{aligned}
Z &= (m - Z_{\dot{w}})\dot{w} + (my_g - Z_{\dot{p}})\dot{p} - (mx_g - Z_{\dot{q}})\dot{q} - (Y_{\dot{v}}v + mv)p \\
&\quad +(X_{\dot{u}}u - (mu))q - (Z_w)w - f_B \cos\theta \cos\phi
\end{aligned} \tag{6.3}
$$

$$
\begin{aligned}
M &= (-mx_g - Z_{\dot{q}})\dot{w} - (I_{xy} - K_{\dot{q}})\dot{p} + (I_{yy} - M_{\dot{q}})\dot{q} \\
&\quad +(Z_{\dot{w}}w + mw)u - (X_{\dot{u}}u + mu)w + (N_{\dot{r}}r - I_{zz}r)p \\
&\quad -(K_{\dot{p}}p - I_x p)r + (M_q)q - Z_B \cos\theta \sin\phi
\end{aligned} \tag{6.4}
$$

$$
\begin{aligned}
N &= (-my_g - X_{\dot{r}})\dot{u} + (mx_g - Y_{\dot{r}})\dot{v} + (I_{zz} - N_{\dot{r}})\dot{r} \\
&\quad -(Y_{\dot{v}}v - mv)u + (X_{\dot{u}}u + mu)v - (M_{\dot{q}}q - I_{yy}q)p \\
&\quad +(K_{\dot{p}}p - I_{xx}p)q - (N_r)r
\end{aligned} \tag{6.5}
$$

where m denote the mass of the vehicle, $X_i, Y_i, Z_i, K_i, M_i, N_i$ are the linear damping coefficients; $i : u, v, w, p, q, r$ y $X_{\dot{k}}, Y_{\dot{k}}, Z_{\dot{k}}, K_{\dot{k}}, M_{\dot{k}}, N_{\dot{k}}$ represent the hydrodynamic coefficients of aggregate mass; $\dot{k} : \dot{u}, \dot{v}, \dot{w}, \dot{p}, \dot{q}, \dot{r}$. In addition, x_g, y_g, z_g are elements of r_g which defines the distance from the origin of the fixed frame to the body O_B to the C_G of the robot, f_B the buoyancy forces, I_{xx}, I_{yy}, I_{zz} are the moments of inertia and I_{xy} it is a product.

Considering that the vehicle moves at low speed and the features contemplated in its design where the roll angle is mechanically stable and, as a consequence $\phi \approx 0$. This will imply that the lateral displacement is also small and could be neglected, hence, $v, \dot{v}, p, \dot{p} \approx 0$. Additionally as the origin of the body structure O_B coincides with the C_G, then $x_g, y_g, z_g \approx 0$, for more details, see [13]. Therefore, it can be rewrite (6.2–6.5) as:

$$\bar{I}_u \dot{u} \approx (k_{wu}q - mr)w + u_x + \delta_x \tag{6.6}$$

$$\bar{I}_z \dot{w} \approx u_z + \bar{I}_u uq + \delta_z \tag{6.7}$$

$$\bar{I}_q \dot{q} \approx u_\theta + \bar{k}_{uw} uw + \delta_\theta \tag{6.8}$$

$$\bar{I}_r \dot{r} \approx u_\psi + \delta_\psi \tag{6.9}$$

where $\bar{I}_u = m - X_{\dot{u}}$, $\bar{I}_z = m - Z_{\dot{w}}$, $\bar{I}_q = I_{yy} - M_{\dot{q}}$, $\bar{I}_r = I_{zz} - N_{\dot{r}}$, $\bar{k}_{uw} = k_{uw} - k_{wu} = X_{\dot{u}} - Z_{\dot{w}}$. For easy reading rename $u_x, u_z, u_\theta, u_\psi$ respectively as the control inputs X, Z, M, N, furthermore $\delta_x, \delta_z, \delta_\theta, \delta_\psi$ can be considered as an unknown and external perturbation or a nonlinear uncertainty in the model that depends on the states of the vehicle.

Using Computational Fluid Dynamics (CFD) software and the 3D CAD design of the vehicle, taking into account the characteristics and materials used in its design, the following parameters are estimated: $m = 33.9\,kg$, $I_{xx} = 0.99\ N * m$, $I_{yy} = 2 \times 10^{-2}\ N * m$, $I_{zz} = 2 \times 10^{-2}\ N * m$, $X_{\dot{u}} = 0.963\,kg$, $Y_{\dot{v}} = 0.367\,kg$, $Z_{\dot{w}} = 0.367\,kg$, $K_{\dot{p}} = 2 \times 10^{-3}\ N * m$, $M_{\dot{q}} = 2 \times 10^{-3}\ N * m$ and $N_{\dot{r}} = 2 \times 10^{-3}\ N * m$.

6.3.2 *Dynamic Model with Respect to the Inertial Frame*

The objective is to validate the strategies, then, it is more common the use of sensors referred to the Euler angles for the orientation and the inertial frame for the position. In [18], we propose the transformation of the angular velocity and the linear velocity of the body frame to the inertial frame, as:

$$\begin{aligned} \dot{\eta_2} &= J_2(\eta_2)\nu_2 \\ \dot{\eta_1} &= J_1(\eta_1)\nu_1 \end{aligned} \tag{6.10}$$

developing the equation we have

$$\begin{aligned} p &= \dot{\phi} - \sin\theta\dot{\psi} \\ q &= \dot{\theta}\cos\phi + \dot{\psi}\cos\theta\sin\phi \\ r &= -\dot{\theta}\sin\phi + \dot{\psi}\cos\theta\cos\phi \\ u &= \dot{x}\cos\psi\cos\theta + \dot{y}\sin\psi\cos\theta - \dot{z}\sin\theta \\ \\ v &= -\dot{x}\sin\psi + \dot{y}\cos\psi \\ w &= \dot{x}\cos\psi\sin\theta + \dot{y}\sin\psi\sin\theta + \dot{z}\cos\theta \end{aligned} \tag{6.11}$$

taking into account that $\phi \approx 0$ and $\dot{\phi} \approx 0$, in addition as the lateral displacement is small $\dot{y} \approx 0$, this implies that:

$$\begin{aligned} p &= -\sin\theta\dot{\psi} \\ q &= \dot{\theta} \\ r &= \dot{\psi}\cos\theta \\ u &= \dot{x}\cos\psi\cos\theta - \dot{z}\sin\theta \\ v &= -\dot{x}\sin\psi \approx 0 \\ w &= \dot{x}\cos\psi\sin\theta + \dot{z}\cos\theta \end{aligned} \tag{6.12}$$

considering that the orientation is stabilized previously, therefore the equations of motion (6.6–6.9) in the body frame can be represented in the inertial frame as:

$$\bar{I}_u\ddot{x} = u_x + \delta_x \tag{6.13}$$

$$\bar{I}_z\ddot{z} = u_z + \bar{I}_u\dot{x}\dot{\theta} + \delta_z \tag{6.14}$$

$$\bar{I}_q\ddot{\theta} = u_\theta + \bar{k}_{uw}uw + \delta_\theta \tag{6.15}$$

$$I_{\psi_1}\ddot{\psi} = u_\psi + I_{\psi_2}\dot{\psi}\dot{\theta} + \delta_\psi \tag{6.16}$$

with $I_{\psi_1} = \bar{I}_r\cos\theta$ y $I_{\psi_2} = \bar{I}_r\sin\theta, u = \dot{x}\cos\psi\cos\theta + \dot{y}\sin\psi\cos\theta - \dot{z}\sin\theta, w = \dot{x}\cos\psi\sin\theta + \dot{y}\sin\psi\sin\theta + \dot{z}\cos\theta$.

6.4 Control of Underwater Vehicle

To validate the performance of the prototype, a nonlinear algorithm with adaptive and integral properties was implemented to stabilize an underwater vehicle. The design and stability analysis are presented in [13], the contribution of this controller in comparison to those in the literature is to add an integral component to obtain robustness in the system against unknown and external perturbations. Additionally, the adaptive property is used to estimate the unknown and non-linear disturbances in the model as the hydrodynamic parameters.

Thus, to stabilize the underwater robot it is proposed to use the following control law to stabilize yaw.

$$u_\psi = I_{\psi_1}(\dot{\rho}^1_\psi - e_\psi - k_{\dot{\bar{e}}_\psi}\dot{\bar{e}}_\psi) - I_{\psi_2}\dot{\psi}\dot{\theta} - \hat{\delta}_\psi \tag{6.17}$$

with $e_\psi = \psi - \psi_d, \rho_\psi^1 = \dot{\psi}_d - k_{s_\psi} s_\psi - k_\psi e_\psi, \dot{\hat{\delta}}_\psi = \frac{1}{\bar{I}_{\psi_1}} k_{\delta_\psi} \dot{\bar{e}}_\psi$ y $\dot{s}_\psi = e_\psi$. In the case of pitch, similarly to the previous case, applying the procedure of the previously proposed controller, we have:

$$u_\theta = \bar{I}_q(\dot{\rho}_\theta^1 - e_\theta - k_{\dot{\bar{e}}_\theta} \dot{\bar{e}}_\theta) - \bar{k}_{uw} uw - \hat{\delta}_\theta \tag{6.18}$$

with $e_\theta = \theta - \theta_d$, $\rho_\theta^1 = \dot{\theta}_d - k_{s_\theta} s_\theta - k_\theta e_\theta$, $\dot{\hat{\delta}}_\theta = \frac{1}{\bar{I}_q} k_{\delta_\theta} \dot{\bar{e}}_\theta$ and $\dot{s}_\theta = e_\theta$.

To stabilize the AUV-UMI in position implies that the orientation were stabilized previously, hence, $\ddot{\psi}, \dot{\psi} \to 0$, $\psi \to \psi_d, \ddot{\theta}, \dot{\theta} \to 0$ and $\theta \to \theta_d$. That means $\cos\theta \neq 0$ y $I_{\psi_1} \neq 0$ en (6.17). Consequently, the controller produces:

$$u_x = \bar{I}_u(\dot{\rho}_x^1 - e_x - k_{\dot{\bar{e}}_x} \dot{\bar{e}}_x) - \hat{\delta}_x \tag{6.19}$$

with $e_x = x - x_d$; $\rho^1 = \dot{x}_d - k_{s_x} s_x - k_x e_x$; $\dot{\hat{\delta}}_x = \frac{1}{\bar{I}_u} k_\delta \dot{\bar{e}}_x$ and $\dot{s}_x = e_x$. Implies that, $\ddot{x}, \dot{x} \to 0$ y $x \to x_d$. Finally, adapting the previously described procedure of the controller for z has:

$$u_z = \bar{I}_z(\dot{\rho}_z^1 - e_z - k_{\dot{\bar{e}}_z} \dot{\bar{e}}_z) - \hat{\delta}_z \tag{6.20}$$

where $e_z = z - z_d$; $\rho^1 = \dot{z}_d - k_{s_z} s_z - k_z e_z$; $\dot{\hat{\delta}}_z = \frac{1}{\bar{I}_z} k_\delta \dot{\bar{e}}_z$ and $\dot{s}_z = e_z$. This implies that $\ddot{z}, \dot{z} \to 0$ and $z \to z_d$.

6.4.1 Numerical Simulation Results

The controllers proposed to stabilize the vehicle were validated with numerical simulations, which illustrate the efficiency of the BIA. The simulation show that the vehicle follow a sinusoidal reference in yaw $\psi_d = (50 * sin(0.1 * t) + 100)$ deg, in the pitch a constant reference $\theta_d = 4$ deg and that maintains a constant depth $z_d = 1.5$ m. It should be noted that a constant perturbation $\delta = 30$ was considered in all cases.

It is important to note that the equations that model vehicle dynamics were used, a MATLAB simulation was created to fully characterize the position and orientation of the vehicle with respect to the inertial frame. Note in Figs. 6.6 and 6.7, that the closed loop system against a constant perturbation remains stable and the disturbance is estimated. The gains used in the yaw simulation are: $k_{s_\psi} = 9 \times 10^{-2}$, $k_\psi = 12$, $k_{\dot{\bar{e}}_\psi} = 15$, $k_{\delta_\psi} = 12 \times 10^{-3}$, with initial conditions $\psi(0) = \dot{\psi}(0) = 0$, for the pitch is: $k_{s_\theta} = 95 \times 10^{-3}$, $k_\theta = 12$, $k_{\dot{\bar{e}}_\theta} = 15$, $k_{\delta_\theta} = 2 \times 10^{-2}$, with initial conditions $\theta(0) = \dot{\theta}(0) = 0$ and with depth $k_{s_z} = 0.9$, $k_z = 8$, $k_{\dot{\bar{e}}_z} = 12$, $k_{\delta_z} = 0.8$, with initial conditions $z(0) = \dot{z}(0) = 0$.

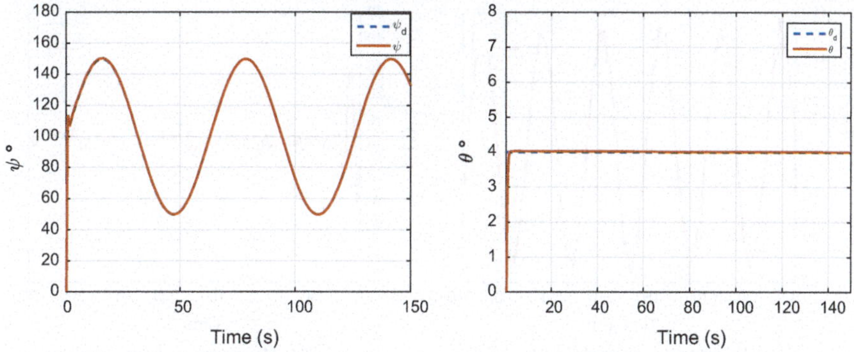

(a) Yaw control response

(b) Pitch control response

Fig. 6.6 Efficiency of the controller to stabilize the orientation considering a constant disturbance

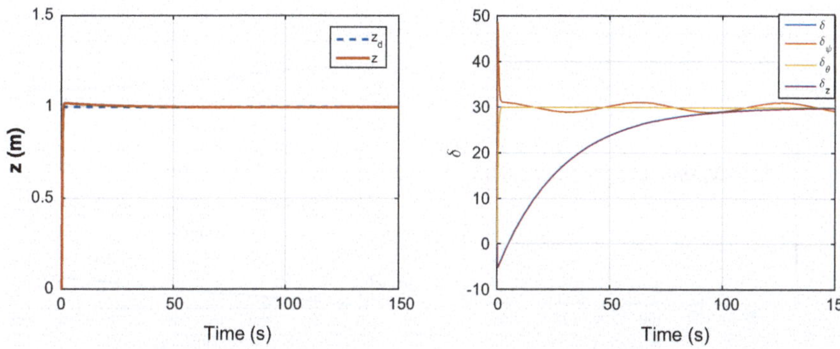

(a) Depth control response

(b) Estimation of disturbance

Fig. 6.7 Efficiency of the controller to stabilize the depth and estimation of disturbances

6.5 Real Time Tests

The response of the BIA controller and the operation of the AUV-UMI robot is validated in real time, for which the proposed control law was implemented. The experiment was carried out in a pool and a transverse stream was generated with the aid of a hydraulic pump in order to produce an external disturbance. The test consists of stabilizing yaw angles, pitch and depth of our vehicle, because the prototype does not yet have a sensor that allows us to measure the position x and its longitudinal velocity. Then, it is proposed that the prototype be submerged to a depth of 1 m with a pitch of $\theta_d = 4$ deg and follow a sinusoidal reference $\psi_d = (50 * sin(0.1 * t) + 100)$ deg at the yaw angle. Later the vehicle is then moved in the longitudinal axis x, in manual mode to increase the hydrodynamic parameters. The gains used in this yaw experiment are: $k_{s_\psi} = 9 \times 10^{-4}$, $k_\psi = 14.11$, $k_{\dot{e}_\psi} = 12.6$, $k_{\delta_\psi} = 15 \times 10^{-6}$, for pitch: $k_{s_\theta} = 989 \times 10^{-7}$, $k_\theta = 14.15$, $k_{\dot{e}_\theta} = 11.75$, $k_{\delta_\theta} = 15 \times 10^{-5}$ and

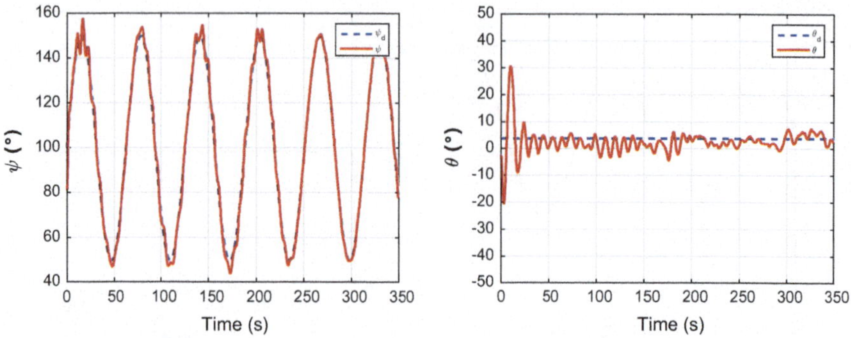

(a) Response of yaw control in experimental test

(b) Response of pitching in experimental test

Fig. 6.8 Efficiency of the controller to stabilize the orientation in experimental test

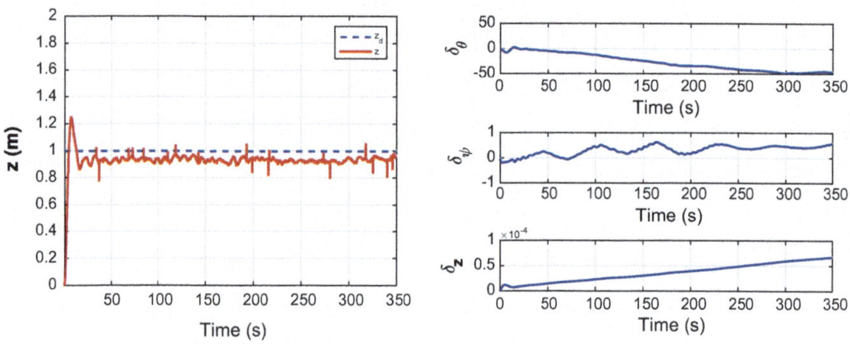

(a) Response of depth control in experimental test

(b) Estimation of disturbances

Fig. 6.9 Efficiency of the controller to stabilize the depth and estimation of the perturbations in experimental test

for the depth z are: $k_{s_z} = 99 \times 10^{-5}$, $k_z = 2$, $k_{\dot{\bar{e}}_z} = 2.5$, $k_{\delta_z} = 11 \times 10^{-5}$. Figures 6.8 and 6.9 shows the performance of the controller in real time, observe here the good behavior of the control algorithms and the convergence towards the reference signal.

Note that the estimated perturbation for δ_ψ decreases as the vehicle tracks the sinusoidal signal, while δ_θ decreases and δ_z remains small while the controller stabilizes the vehicle.

6.6 Conclusion

The development of the AUV-UMI, a modular vehicle, capable of operating in ROV or AUV mode and acquiring information about its environment and its own status data was described. In order to validate theoretically and experimentally The proposed controller, obtained and analyzed the dynamic model of the underwater robot contemplating its characteristics and limitations with the objective of implementing a nonlinear control algorithm with the robustness necessary to stabilize the vehicle, several simulations were performed that illustrate the good operation of the system In closed loop even in the presence of unknown and external disturbances. Experimental results verify the behavior of the BIA by stabilizing yaw, pitch, and depth states.

Future work includes implementing the vehicle with more sensors such as: sonar, echo sounder or DVL to measure the longitudinal axis, with the intention of performing more complex tasks autonomously. In addition, it is proposed the comparative analysis between proposed control law and other well established control technique in order to quantify the performance and results of presented control approach.

References

1. A. Alvarez, A. Caffaz, A. Caiti, G. Casalino, L. Gualdesi, A. Turetta, R. Viviani, Folaga: a low-cost autonomous underwater vehicle combining glider and AUV capabilities. Ocean Eng. **36**, 24–38 (Elsevier, 2009)
2. M. Aras, M. Shahrieel, H.A. Kasdirin, M.H. Jamaluddin, M.F. Basar, *Design and Development of an Autonomous Underwater Vehicle (AUV-FKEUTeM)* (Universiti Malaysia Pahang, 2009)
3. E. Cavallo, R.C. Michelini, V.F. Filaretov, Conceptual design of an AUV equipped with a three degrees of freedom vectored thruster. J. Intell. Robot. Syst. **39**, 365–391 (Springer, 2004). (3rd ed)
4. K.D. Do, Z.P. Jiang, J. Pan, H. Nijmeijer, A global output-feedback controller for stabilization and tracking of underactuated ODIN: a spherical underwater vehicle. Automatica **40**, 117–124 (Elsevier, 2004)
5. B.M. Ferreira, J. Jouffroy, A.C. Matos, N.A Cruz, Control and guidance of a hovering AUV pitching up or down, in *Oceans* (IEEE, 2012), pp. 1–7
6. A.S. Gadre, J.J. Mach, D.J. Stilwell, C.E. Wick, Design of a prototype miniature autonomous underwater vehicle, in Proceedings 2003 IEEE/RSJ International Conference on Intelligent Robots and Systems, 2003 (IROS 2003), vol. 1, 3rd edn. (IEEE, 2003), pp. 842–846
7. J.G. Graver, *Underwater Gliders: Dynamics, Control and Design* (Citeseer, 2005)
8. J. Guo, F-C. Chiu, C-C. Huang, Design of a sliding mode fuzzy controller for the guidance and control of an autonomous underwater vehicle. Ocean Eng. **30**(16), 2137–2155 (Elsevier, 2003)
9. L. Lapierre, D. Soetanto, Nonlinear path-following control of an AUV. Ocean Eng. **34**(11), (Elsevier, 2007)
10. J.H. Li, P.M. Lee, Design of an adaptive nonlinear controller for depth control of an autonomous underwater vehicle. Ocean Eng. **32**(17), 2165–2181 (Elsevier, 2005)
11. S. Licht, V. Polidoro, M. Flores F.S. Hover, M.S. Triantafyllou, Design and projected performance of a flapping foil AUV. IEEE J. Ocean. Eng. **29**, 786–794 (IEEE, 2004). 3rd ed

12. D. Maalouf, I. Tamanaja, E. Campos, A. Chemori, V. Creuze, J. Torres, R. Lozano, From pd to nonlinear adaptive depth-control of a tethered autonomous underwater vehicle. IFAC Proc. Vol. **46**(2), 743–7480 (Elsevier, 2013)
13. A. Manzanilla, P.Castillo, R. Lozano, Nonlinear algorithm with adaptive properties to stabilize an underwater vehicle: real-time experiments, in *IFAC-PapersOnLine* (Elsevier, 2017)
14. V.K. Pshikhopov, M.Y. Medvedev, A.R. Gaiduk, B.V. Gurenko, Control system design for autonomous underwater vehicle. in *Robotics Symposium and Competition (LARS/LARC), 2013 Latin American* (IEEE, 2013), pp. 77–82
15. D. Ribas, P. Ridao, L. Magí, N. Palomeras, M. Carreras, The Girona 500, a multipurpose autonomous underwater vehicle. in *OCEANS, 2011 IEEE-Spain* (2011), pp. 1–5
16. J.L. Rosendo, B. Clement, F. Garelli, Sliding mode reference conditioning for path following applied to an AUV. IFAC-PapersOnLine **49**(23) (Elsevier, 2016)
17. SNAME (1952), Nomenclature for treating the motion of a submerged body through a fluid jr. Technical and Research Bulletin No. 1–5 (New York, 1952)
18. I.F. Thor, *Guidance and Control of Ocean Vehicles* (Wiley, 1994)
19. I.F. Thor, *Handbook of Marine Craft Hydrodynamics and Motion Control* (Wiley, 2011)
20. I. Torres, Iván, J. Torres, R. Lozano, A new AUV configuration with four tilting thrusters: navigation and hover tasks. in *2012 9th International Conference on Electrical Engineering, Computing Science and Automatic Control (CCE)* (IEEE, 2012), pp. 1–6
21. L. Wang, L.J. Zhang, H.M. Jia, H.B. Wang, Horizontal tracking control for AUV based on nonlinear sliding mode. in *2012 International Conference on Information and Automation (ICIA)* (IEEE, 2012), pp. 460–463
22. S.A. Watson, D.J.P. Crutchley, P.N. Green, The design and technical challenges of a micro-autonomous underwater vehicle. in *2011 International Conference on Mechatronics and Automation (ICMA)* (2011), pp. 567–572

Chapter 7
Secure a Zone from Intruders with a Group Robots

Khadimoullah Vencatasamy, Luc Jaulin and Benoît Zerr

Abstract The paper proposes new method to secure a zone of the world from intruders using a group a robots. The principle is to control the group in order to form a chain through which no intruder can go without being detected. We use a set membership method based on interval analysis and mathematical morphology to guarantee that no intruder is inside the secure zone. The approach is illustrated by an example where the environment is the Bay of Biscay, the intruder is a submarine and the group of robots consists of small underwater autonomous vehicles.

7.1 Introduction

We consider n robots $\mathcal{R}_1, \ldots, \mathcal{R}_n$ at positions $\mathbf{a}_1, \ldots, \mathbf{a}_n$ and moving in a 2D world [1]. Each robot has a visibility zone. If an intruder is inside the visibility zone of one robot, it is detected. The robots have to collaborate to guarantee that there is no moving intruder inside a subzone of a compact subset $\mathbb{O}$ of $\mathbb{R}^2$, representing the world.

This project is motivated by the detection of submarine intruders inside the Bay of Biscay (golfe de Gascogne) by low cost robots [2] equipped with low cost and passive sensors (microphones, cameras [3], electric sense [4, 5]). The set $\mathbb{O}$ corresponds to the ocean and the intruder is a submarine considered as an enemy. We will avoid using expensive sonars such as side-scan sonars [6] which consume energy and which are not discrete.

For simplicity, we consider a 2D environment, and the robots are able find their position with a given accuracy using a state observer. We assume that each robot is

K. Vencatasamy · L. Jaulin (✉) · B. Zerr
Lab-STICC, UMR CNRS 6285, ENSTA Bretagne, Brest, France
e-mail: luc.jaulin@ensta-bretagne.fr

K. Vencatasamy
e-mail: khadimoullah.vencatasamy@ensta-bretagne.fr

B. Zerr
e-mail: benoit.zerr@ensta-bretagne.fr

L. Jaulin et al. (eds.), *Marine Robotics and Applications*, Ocean Engineering
& Oceanography 10, https://doi.org/10.1007/978-3-319-70724-2_7

able to detect any intruder inside a disk of a known radius. Moreover, the intruder's dynamics is assumed to be known (but not its control).

The goal of this paper is twofold: (1) to characterize a set for which we can guarantee that there is no intruder (this corresponds to the secure zone) and (2) to find a control strategy for the group of robots in order to extend the secure zone as much as possible. For this purpose, we will use interval analysis and control theory.

Set-membership approaches have mainly been used in the context of underwater robotics [7–9] for guaranteed localization and mapping. For the first time, we use it here to guarantee that a zone is secured.

In this paper, we are not interested in a dynamical control of one robot in order to perform a particular mission, but more by a collective control of a group of robots for a common task.

The paper is organized as follows. Section 7.2 recalls how to control independently a group of robots that cannot communicate. Section 7.3 shows how to compute an inner and outer approximation of the secure zone. An illustration of our approach is given in Sect. 7.4 where a group of robots collaborate to secure the Bay of Biscay. Section 7.5 concludes this paper and proposes some perspectives.

7.2 Control of a Group of Robots

The first step to be solved to secure a zone is to control a group of robots assuming that the robots cannot communicate, in order to be as silent as possible. In such a case, classical linear control [10] do not apply. The general method in this situation is the flocking approach [11] or to consider that all trajectories of all robots have been decided a priori.

Remark. When we say that the robots do not communicate, it means that they do not share any information stored inside their memory. But, we can still assume that each robot, via its own sensors are able estimate the state (position, orientation and speed) of its neighbors. We do not consider the perception of neighbors as a communication. Now, since the behavior of each robot is influenced by what is stored in its memory, a robot can guess some part of the memory of its neighbors by analyzing their behaviors. Moreover, the robots can voluntarily change their behaviors in order to transmit a coded message. It is what insects do when they perform some kind of dance to communicate. For instance, by performing the figure-eight dance, the honey bees can share with other members of the colony, information about the direction and distance to flowers yielding nectar and pollen, to water sources, or to new nest-site location. In such a case, there is a clear intention to communicate. Here, we shall consider that we have no communication when there is no intention to transmit any message from one robot to another robot.

7.2.1 Flocking

To illustrate the flocking, let us consider m robots described by the following state equations:

$$\begin{cases} \dot{x}_i = \cos\theta_i \\ \dot{y}_i = \sin\theta_i \\ \dot{\theta}_i = u_i \end{cases}$$

The state vector is $\mathbf{x}(i) = (x_i, y_i, \theta_i)$. In this model, the speed in always equal to 1, which means that each robot is only control its heading through the input u_i. These robots can see all other robots, but are not able to communicate with them. We want that these robots behave as a flock as illustrated by Fig. 7.1. Basic models of flocking behaviors are controlled by three rules of Reynolds:

- The *separation* corresponds to a short range repulsion force. This allows the flock to avoid crowding neighbors.
- The *alignment* force implies that each robot steers towards average heading of its neighbors.
- The *cohesion* is a long range attraction where each robot of the flock is attracted by its neighbors.

Using a artificial potential field method [12] (see also [13] or [14] in the context of ocean robotics), we can find a controller for each robot to obtain a flock. In Fig. 7.1, the robots are first initialized randomly in the same zone of the world. Then, following the rules of Reynolds, they adopt the behavior of a swarm.

For this, we define the potential of the ith robot as

$$V_i = \sum_{j \neq i} \alpha \left\| \mathbf{p}(i) - \mathbf{p}(j) \right\|^2 + \frac{\beta}{\left\| \mathbf{p}(i) - \mathbf{p}(j) \right\|}$$

where $\mathbf{p}(i) = \left(x_i,\ y_j\right)$. The quadratic part of the sum corresponds to the cohesion and the hyperbolic term corresponds to the separation. The coefficients α and β have to be tuned. For instance, if β is high, the repulsion will be stronger and the flock will be more scattered. The gradient of V_i is

$$\frac{dV_i}{d\mathbf{p}(i)} = \sum_{j \neq i} \left(2\alpha \left(\mathbf{p}(i) - \mathbf{p}(j)\right) - \frac{\beta \left(\mathbf{p}(i) - \mathbf{p}(j)\right)^T}{\left\| \mathbf{p}(i) - \mathbf{p}(j) \right\|^3} \right).$$

Since we want to decrease V_i, we have to follow the opposite direction of the gradient. Moreover, in order to follow the alignment rule of Reynolds each robot should get the same heading as the other robots. Thus, the controller is built in order to follow the direction

$$\sum_{j \neq i} \left(-2\alpha \left(\mathbf{p}(i) - \mathbf{p}(j)\right) - \frac{\beta \left(\mathbf{p}(i) - \mathbf{p}(j)\right)^T}{\left\| \mathbf{p}(i) - \mathbf{p}(j) \right\|^3} + \begin{pmatrix} \cos\theta_j \\ \sin\theta_j \end{pmatrix} \right).$$

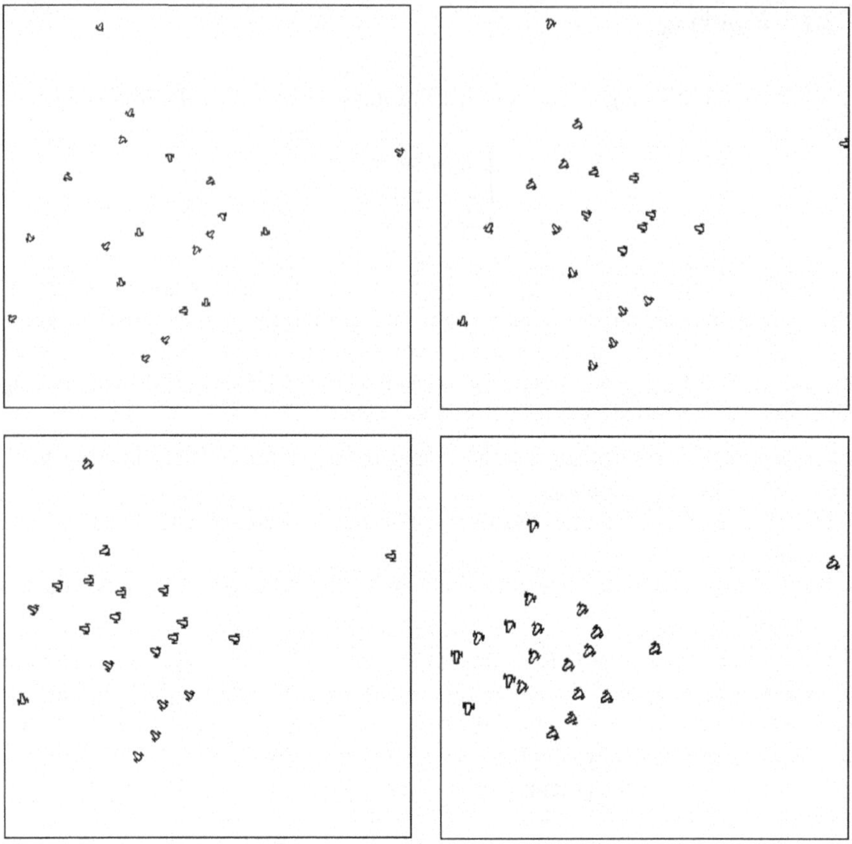

Fig. 7.1 Illustration of the flocking behavior from a random initialization in the case where $m = 20$

The flocking is a nice method to move a group of robots and could be used to find an intruder. Moreover, no communication or distributed computation is needed. Now we cannot guarantee anything with flocking behavior about the stability or the cohesion (see e.g., [1]). We need a control which is more deterministic as proposed in the following subsection.

7.2.2 Group of Robots Forming a Chain

To secure a zone, with guarantee, the first idea that comes to mind is to build a virtual chain made with robots. If the chain is built with the right spacing and takes into account the visibility zone of each robot, the intruder will not be able to cross the chain without being detected. Moreover, we want the robots move with respect to a cycle so that they can be able to recharge their batteries at some points of the cycle.

To illustrate how this can be done, assume that we have a group of $m = 20$ robots the motion of which is described by the state equation

$$\begin{cases} \dot{x}_1 = x_4 \cos x_3 \\ \dot{x}_2 = x_4 \sin x_3 \\ \dot{x}_3 = u_1 \\ \dot{x}_4 = u_2 \end{cases}$$

where (x_1, x_2) corresponds to the position of the robot, x_3 to its heading and x_4 to its speed. We want the robots to follow an ellipse which corresponds to a barrier for the intruder.

Circle. Assume first that this ellipse is a circle. We want a controller for each of these robots so that the ith robot follows the trajectory

$$\begin{pmatrix} \cos(at + \frac{2i\pi}{m}) \\ \sin(at + \frac{2i\pi}{m}) \end{pmatrix},$$

where $a = 0.1$. As a consequence, after the initialization step, all robots are uniformly distributed on the unit circle, turning around the origin. To do this, we can use a feedback linearization method taking as an output $\mathbf{y} = (x_1, x_2)$. We get the controller

$$\mathbf{u} = \begin{pmatrix} -x_4 \sin x_3 & \cos x_3 \\ x_4 \cos x_3 & \sin x_3 \end{pmatrix}^{-1} \cdot \left(\mathbf{c}\,(t) - \begin{pmatrix} x_1 \\ x_2 \end{pmatrix} + 2\dot{\mathbf{c}}\,(t) - 2 \begin{pmatrix} x_4 \cdot \cos(x_3) \\ x_4 \cdot \sin(x_3) \end{pmatrix} + \ddot{\mathbf{c}}\,(t) \right), \quad (7.1)$$

where $\mathbf{c}(t)$ is the desired position. For the ith robot, we take

$$\mathbf{c}\,(t) = \begin{pmatrix} \cos(at + \frac{2i\pi}{m}) \\ \sin(at + \frac{2i\pi}{m}) \end{pmatrix}, \dot{\mathbf{c}}\,(t) = a \cdot \begin{pmatrix} -\sin(at + \frac{2i\pi}{m}) \\ \cos(at + \frac{2i\pi}{m}) \end{pmatrix}, \ddot{\mathbf{c}}\,(t) = -a^2 \mathbf{c}\,(t) \cdot$$

Ellipse. By using a linear transformation of the unit circle, we can change the controllers for the robots so that all robots stay on a moving ellipse. For instance, assume that we want a first axis of length $20 + 15 \cdot \sin(at)$ and the second axis of length 20. Moreover, we want the ellipse rotating by choosing an angle for the first axis of $\theta = at$.

To get the right ellipse, for each $\mathbf{c}\,(t)$, we apply the transformation

$$\mathbf{w}\,(t) = \underbrace{\begin{pmatrix} \cos\theta & -\sin\theta \\ \sin\theta & \cos\theta \end{pmatrix}}_{=\mathbf{R}} \cdot \underbrace{\begin{pmatrix} 20 + 15 \cdot \sin(at) & 0 \\ 0 & 20 \end{pmatrix}}_{=\mathbf{D}} \cdot \mathbf{c}\,(t)$$

where $\mathbf{w}(t)$ is the new desired position. To apply our controller, we also need the two first derivatives of $\mathbf{w}(t)$. We have

$$\dot{\mathbf{w}} = \mathbf{R} \cdot \mathbf{D} \cdot \dot{\mathbf{c}} + \mathbf{R} \cdot \dot{\mathbf{D}} \cdot \mathbf{c} + \dot{\mathbf{R}} \cdot \mathbf{D} \cdot \mathbf{c},$$

where

$$\dot{\mathbf{D}} = \begin{pmatrix} 15a \cdot \cos(at) & 0 \\ 0 & 0 \end{pmatrix}, \text{ and } \dot{\mathbf{R}} = a \cdot \begin{pmatrix} -\sin\theta & -\cos\theta \\ \cos\theta & -\sin\theta \end{pmatrix}.$$

Moreover

$$\ddot{\mathbf{w}} = \mathbf{R} \cdot \mathbf{D} \cdot \ddot{\mathbf{c}} + \mathbf{R} \cdot \ddot{\mathbf{D}} \cdot \mathbf{c} + \ddot{\mathbf{R}} \cdot \mathbf{D} \cdot \mathbf{c} + 2 \cdot \dot{\mathbf{R}} \cdot \mathbf{D} \cdot \dot{\mathbf{c}} + 2 \cdot \mathbf{R} \cdot \dot{\mathbf{D}} \cdot \dot{\mathbf{c}} + 2 \cdot \dot{\mathbf{R}} \cdot \dot{\mathbf{D}} \cdot \mathbf{c}$$

where

$$\ddot{\mathbf{D}} = \begin{pmatrix} -15a^2 \cdot \sin(at) & 0 \\ 0 & 0 \end{pmatrix} \text{ and } \ddot{\mathbf{R}} = -a^2 \cdot \mathbf{R}.$$

We can now apply the controller given by Eq. (7.1), where $\mathbf{c}(t)$ is now replaced by $\mathbf{w}(t)$. Fig. 7.2 illustrates the behavior of the control law and shows that the group moves exactly with the ellipse.

We have shown how a group of robots can follow an ellipse. As a result, we may interpret the group as a unique *ellipse robot* which can have an arbitrary shape. Now, when the ellipse becomes flat (as it is the case when we want to build a chain for the intruders), the placement of the robots is not uniform: the robots concentrate around the two apogee points. The next subsection explains how to get a uniform distribution of robots along the ellipse.

7.2.3 Convoy

To illustrate how to have a uniform distribution along the ellipse (with a constant distance between robots), consider one robot $\mathcal{R}_A$ described by the following state equations:

$$\begin{cases} \dot{x}_a = v_a \cos\theta_a \\ \dot{y}_a = v_a \sin\theta_a \\ \dot{\theta}_a = u_{a1} \\ \dot{v}_a = u_{a2} \end{cases}$$

where v_a is the speed of $\mathcal{R}_A$ the robot, θ_a its orientation and (x_a, y_a) the coordinates of its center. The robot can be considered as a leader and all robots have to follow it, maintaining a constant distance between two successors. For instance, as previously, we can ask all robots to follow $\mathcal{R}_A$. Therefore, we ask $\mathcal{R}_A$ to follow the trajectory:

$$\begin{cases} \hat{x}_a(t) = L_x \sin(\omega t) \\ \hat{y}_a(t) = L_y \cos(\omega t) \end{cases}$$

with $\omega = 0.1$, $L_x = 20$ and $L_y = 5$. We can take:

$$\mathbf{u}_a = \begin{pmatrix} -v_a \sin\theta_a & \cos\theta_a \\ v_a \cos\theta_a & \sin\theta_a \end{pmatrix}^{-1} \cdot \left(\begin{pmatrix} \hat{x}_a \\ \hat{y}_a \end{pmatrix} - \begin{pmatrix} x_a \\ y_a \end{pmatrix} + \frac{d}{dt} \begin{pmatrix} \hat{x}_a \\ \hat{y}_a \end{pmatrix} - \begin{pmatrix} v_a \cos\theta_a \\ v_a \sin\theta_a \end{pmatrix} \right)$$

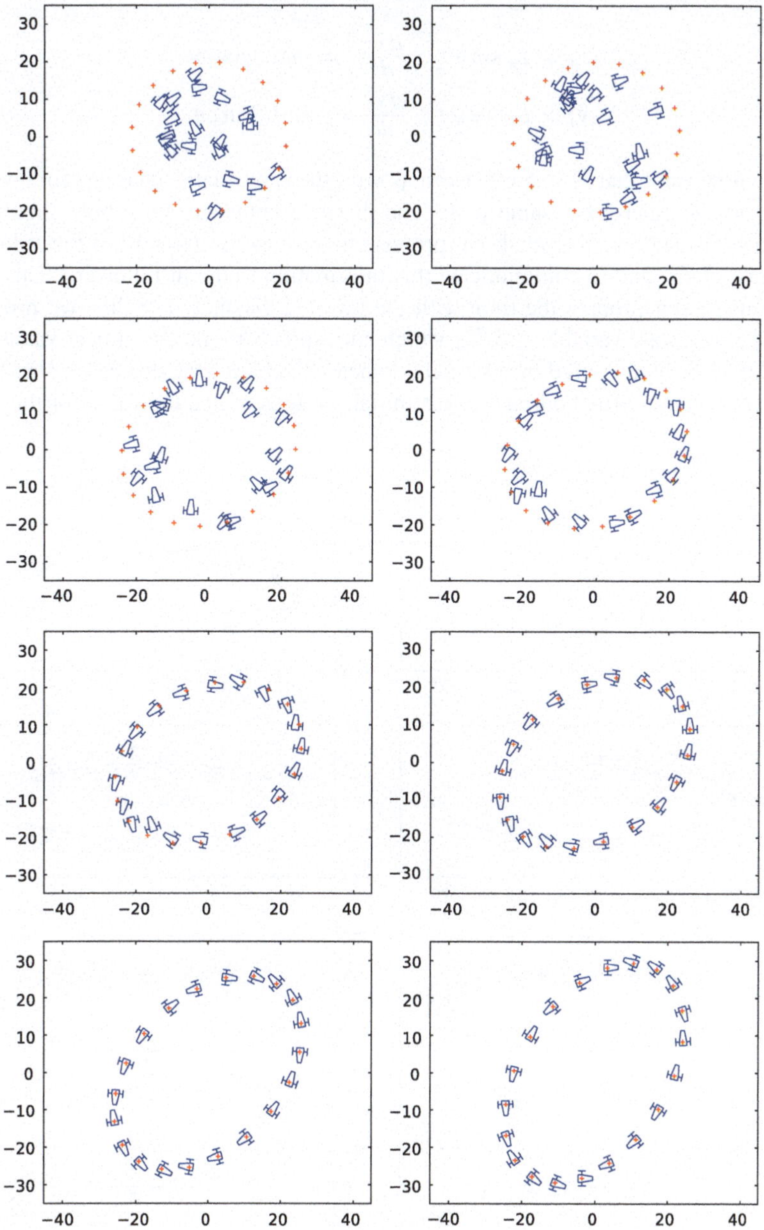

Fig. 7.2 During the transient phase, all robots go toward their setpoints in the ellipse. Then, the group follows exactly the deformation of the ellipse

where

$$\hat{x}_a = L_x \sin \omega t \ , \ \frac{d\hat{x}_a}{dt} = \omega L_x \cos \omega t$$
$$\hat{y}_a = L_y \cos \omega t \ , \ \frac{d\hat{y}_a}{dt} = -\ \omega L_y \sin \omega t.$$

We now want that $m = 6$ other robots with the same state equations follow this robot taking exactly the same path. The distance between two robots should be $d = 5$ m. To achieve this goal, we propose to save every $ds = 0.1$ m the value of the state of $\mathcal{R}_A$ and to communicate this information to the m followers. This will allow us to synchronize the time with the traveled distance. For this, we propose to add a new state variable s to $\mathcal{R}_A$ which corresponds to the curvilinear value that could have been measured by a virtual odometer. Each time the distance ds has been measured by the virtual odometer, s is initialized to zero and the value for the state

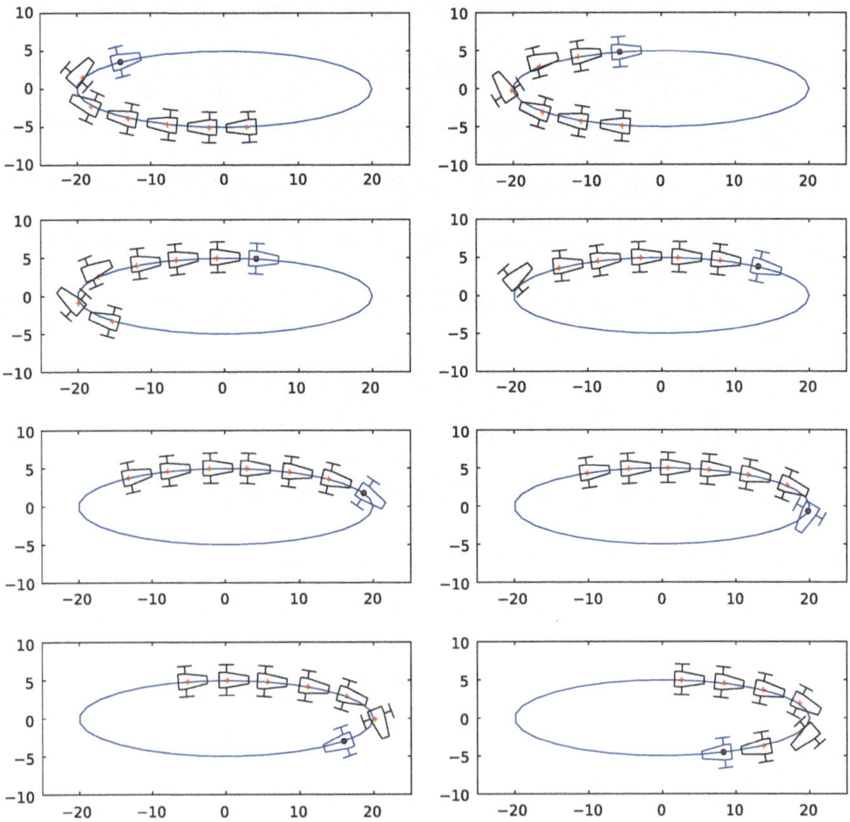

Fig. 7.3 After the transient phase, all robots follow the leader with constant separation distance between two robots

of $\mathcal{R}_A$ is broadcast. We have the new state equation $\dot{s} = v$ which corresponds to the virtual odometer. Each time s is larger than $ds = 0.1$m, we set $s = 0$ and we store the corresponding state of $\mathcal{R}_A$ in a matrix $\mathbf{S}$ which is available to the followers. Since the robot has to be at a distance $d \cdot i$ to $\mathcal{R}_A$, it has to follow the position that $\mathcal{R}_A$ had at a time $t - \delta_i$ when it was at a distance $d \cdot i$ m earlier. Now, the speed to be followed is that of $\mathcal{R}_A$ at the current time t. Indeed, if $\mathcal{R}_A$ slows down, all followers should also slow down immediately. Thus the control to be given at time t is

$$\mathbf{u}(i) = \begin{pmatrix} -v_i \sin\theta_i & \cos\theta_i \\ v_i \cos\theta_i & \sin\theta_i \end{pmatrix}^{-1} \cdot \left(\begin{pmatrix} x_a(t-\delta_i) \\ y_a(t-\delta_i) \end{pmatrix} - \begin{pmatrix} x_i \\ y_i \end{pmatrix} + v_a(t) \cdot \begin{pmatrix} \cos(\theta_a(t-\delta_i)) \\ \sin(\theta_a(t-\delta_i)) \end{pmatrix} - \begin{pmatrix} v_i \cos\theta_i \\ v_i \sin\theta_i \end{pmatrix} \right)$$

where $(x_i, y_i, \theta_i, v_i)$ is the state of the ith robot at time t. The values $(x_a(t-\delta_i), y_a(t-\delta_i), \theta_a(t-\delta_i))$ have been stored in the matrix $\mathbf{S}$. It was $\frac{d \cdot i}{ds}$ steps earlier. Figure 7.3 illustrates the behavior of the control law and shows that the distance between two robots does not depend on the position of the robots on the ellipse.

7.3 Computing the Secure Zone

We consider n robots $\mathcal{R}_1, \ldots, \mathcal{R}_n$ at positions $\mathbf{a}_1, \ldots, \mathbf{a}_n$ in a 2D world $\mathbb{O}$. We assume that this group of robots move in order to build a barrier to protect a zone against the presence of any intruder, for instance, using the techniques presented in the previous section. We now want to provide a numerical set-membership method to guarantee that indeed there is no moving intruder which has entered inside a subzone of $\mathbb{O}$.

7.3.1 Complementary Approach

We show here that the problem of computing the secure zone is equivalent (or more precisely complementary) to building a set-membership observer. For this purpose, we assume the following

- There exists a virtual intruder moving inside $\mathbb{O}$.
- The virtual intruder satisfying the differential inclusion [15]

$$\dot{\mathbf{x}}(t) \in \mathbb{F}(\mathbf{x}(t)),$$

  where $\mathbf{x}(t)$ is its state vector.
- Each robot $\mathcal{R}_i$ has a visibility zone of the form $g_{\mathbf{a}_i}^{-1}([0, d])$ where d is the scope.

Our contribution is to show that characterizing the secure zone translates into a set-membership set estimation problem [16] where $\mathbf{x}(t)$ is shown to be inside the set

$\mathbb{X}(t)$ returned by our set-membership observer. Then we conclude that $\mathbf{x}(t)$ cannot be inside the *complementary* of $\mathbb{X}(t)$. The complementary corresponds to the secure zone. This result can be formalized by the following theorem.

Theorem. The virtual intruder has a state vector $\mathbf{x}(t)$ inside the set

$$\mathbb{X}(t) = \mathbb{O} \cap (\mathbb{X}(t - dt) + dt \cdot \mathbb{F}(\mathbb{X}(t - dt)) \cap \bigcap_i g^{-1}_{\mathbf{a}_i(t)}([d(t), \infty]), \tag{7.2}$$

where $\mathbb{X}(0) = \mathbb{O}$. As a consequence, the secure zone is

$$\mathbb{S}(t) = \overline{proj_{world}(\mathbb{X}(t))}. \tag{7.3}$$

Proof (sketch). Two cases should be considered. If no actual intruder exists then $\mathbb{S}(t)$ cannot contain the intruder. If the virtual intruder is a real one, its state $\mathbf{x}(t)$ is inside $\mathbb{X}(t)$ and its position (which is a part of the state) is inside $proj_{world}(\mathbb{X}(t))$. In both situations, the intruder cannot be inside $\mathbb{S}(t)$.

7.3.2 Mathematical Morphology

For the implementation of Eq. 7.2. We assume that the state of the intruder corresponds to its position. Thus the state space is of dimension 2 and corresponds to the 2D world. As a consequence, the projection operator in (7.3) is not needed anymore and $\mathbb{S}(t) = \overline{\mathbb{X}(t)}$. The set-membership evolution equation can be obtained using mathematical morphology [17] that is now recalled.

The Minkowski sum and the Minkowski difference of two sets $\mathbb{A}$ and $\mathbb{B}$ are defined by

$$\begin{aligned} \mathbb{A} + \mathbb{B} &= \{\mathbf{a} + \mathbf{b} \in \mathbb{R}^n | \mathbf{a} \in \mathbb{A} \text{ and } \mathbf{b} \in \mathbb{B}\} \\ \mathbb{A} - \mathbb{B} &= \{\mathbf{a} - \mathbf{b} \in \mathbb{R}^n | \mathbf{a} \in \mathbb{A} \text{ and } \mathbf{b} \in \mathbb{B}\}. \end{aligned} \tag{7.4}$$

These operations correspond to the addition and difference used in interval computation. When $\mathbb{A}$ or $\mathbb{B}$ are singletons, we get these rules yield:

$$\begin{aligned} \mathbb{A} + \mathbf{b} &= \{\mathbf{a} + \mathbf{b} \in \mathbb{R}^n | \mathbf{a} \in \mathbb{A}\} && \text{(translation by } \mathbf{b}) \\ \mathbb{A} - \mathbf{b} &= \{\mathbf{a} - \mathbf{b} \in \mathbb{R}^n | \mathbf{a} \in \mathbb{A}\} && \text{(translation by } -\mathbf{b}) \\ -\mathbb{A} &= \{-\mathbf{a} \in \mathbb{R}^n | \mathbf{a} \in \mathbb{A}\}. && \text{(symmetry)} \end{aligned} \tag{7.5}$$

Dilatation. The *dilatation* of a set $\mathbb{A}$ by the *structuring element* $\mathbb{B}$ is defined by $\mathbb{A} + \mathbb{B}$.

Erosion. The *erosion* of $\mathbb{A}$ by $\mathbb{B}$ is defined by the three following equivalent relations:

$$\begin{aligned} \mathbb{A} \ominus \mathbb{B} &= \{\mathbf{z} \in \mathbb{R}^n | \mathbb{B} + \mathbf{z} \subseteq \mathbb{A}\} \\ &= \bigcap_{\mathbf{b} \in \mathbb{B}} \mathbb{A} - \mathbf{b} \\ &= \overline{\overline{\mathbb{A}} - \mathbb{B}}. \end{aligned} \tag{7.6}$$

where $\overline{\mathbb{X}}$ denotes the complement of $\mathbb{X}$ relative to $\mathbb{R}^n$.

Opening. The *opening* of $\mathbb{A}$ by $\mathbb{B}$ is obtained by the erosion of $\mathbb{A}$ by $\mathbb{B}$, followed by dilation by $\mathbb{B}$:

$$\mathbb{A} \circ \mathbb{B} = (\mathbb{A} \ominus \mathbb{B}) + \mathbb{B} \tag{7.7}$$

Closing. The *closing* of $\mathbb{A}$ by $\mathbb{B}$ is obtained by the dilatation of $\mathbb{A}$ by $\mathbb{B}$, followed by erosion by $\mathbb{B}$:

$$\mathbb{A} \bullet \mathbb{B} = (\mathbb{A} + \mathbb{B}) \ominus \mathbb{B} \tag{7.8}$$

Properties of the basic operators. Here are some properties of the basic binary morphological operators:

They are increasing, that is, if

$$\begin{array}{lll} \mathbb{A} \subset \mathbb{C} & \Rightarrow & \mathbb{A} + \mathbb{B} \subset \mathbb{C} + \mathbb{B} \\ \mathbb{A} \subset \mathbb{C} & \Rightarrow & \mathbb{A} \ominus \mathbb{B} \subset \mathbb{C} \ominus \mathbb{B} \\ \mathbf{0} \in \mathbb{B} & \Rightarrow & \mathbb{A} \ominus \mathbb{B} \subseteq \mathbb{A} \circ \mathbb{B} \subseteq \mathbb{A} \subseteq \mathbb{A} \bullet \mathbb{B} \subseteq \mathbb{A} + \mathbb{B} \\ \mathbb{A} \subseteq (\mathbb{C} \ominus \mathbb{B}) & \Leftrightarrow & (\mathbb{A} \oplus \mathbb{B}) \subseteq \mathbb{C} \end{array} \tag{7.9}$$

Moreover

$$\begin{aligned} \mathbb{A} + \mathbb{B} &= \mathbb{B} + \mathbb{A} \\ (\mathbb{A} + \mathbb{B}) + \mathbb{C} &= \mathbb{A} + (\mathbb{B} + \mathbb{C}) \\ (\mathbb{A} \ominus \mathbb{B}) \ominus \mathbb{C} &= \mathbb{A} \ominus (\mathbb{B} + \mathbb{C}) \\ \mathbb{A} + \mathbb{B} &= \overline{\overline{\mathbb{A}} \ominus -\mathbb{B}} \\ \mathbb{A} \bullet \mathbb{B} &= \overline{\overline{\mathbb{A}} \circ -\mathbb{B}} \end{aligned} \tag{7.10}$$

7.4 Illustration

To illustrate our method, we consider that there exists an intruder in the set $\mathbb{O}$ which corresponds to the Bay of Biscay with an area of about 220 000 km^2. The set $\mathbb{O}$ is represented by Fig. 7.4. All computations made in this section have been performed using contractor/separator algebra [18, 19], except for the erosion and the dilatation that are taken from OpenCV.

Fig. 7.4 In blue is represented the set $\mathbb{O}$ which is supposed to enclose the virtual intruder

7.4.1 Computation of Secure Zone

Assume that each robot $\mathcal{R}_i$ is able to detect the intruder if the distance is less than a scope distance d. This means that a robot located at position $\mathbf{a}$ and equipped with a range sensor within a scope distance of d, detects that no intruder exists in the set $g_{\mathbf{a}}^{-1}([d,\infty])$, where

$$g_{\mathbf{a}}(\mathbf{x}) = \|\mathbf{x}-\mathbf{a}\| \notin [0,d]. \tag{7.11}$$

Figure 7.5 considers a situation with two robots. The blue zone of all sub-figures is proved to contain the intruder. It corresponds to the right term in (7.2). The first sub-figure corresponds to the static case, i.e., it is given by the set:

$$\mathbb{O} \cap \bigcap_{i\in\{1,2\}} g_{\mathbf{a}_i}^{-1}([d,\infty]). \tag{7.12}$$

Assume now that the virtual intruder has a speed less than $\bar{v}$, i.e., $\|\dot{\mathbf{x}}\| \leq \bar{v}$. The motion of the virtual intruder thus obeys to the following state equation:

$$\begin{cases} \dot{x}_1 = v\cos\psi \\ \dot{x}_2 = v\sin\psi \end{cases} \tag{7.13}$$

where $v \leq \bar{v}$. This also means that, $\mathbb{F}(\mathbf{x})$ corresponds to a disk $\mathbb{D}(\mathbf{0},\bar{v})$ of center $\mathbf{0}$ and radius $\bar{v}$ and does not depend on $\mathbf{x}$. Again, since we assume that $\mathcal{R}_i$ detects nothing, we have $\|\mathbf{x}-\mathbf{a}_i\| > d$. For the initialization, we assume that $\mathbb{X}(0) \subset \mathbb{O} \subset [\mathbf{x}](0)$

Fig. 7.5 No intruder can be inside the magenta zone

where $[\mathbf{x}](0)$ is a box. Sub-Figures (b),(c),(d) of Fig. 7.5 shows the corresponding secure zone painted magenta. The blue zone, contains the virtual intruder.

7.4.2 *Smoother*

It is be possible to take into account the future to increase the secure zone. Indeed, it the intruder is at position $\mathbf{x}$, and if for all feasible maneuvers, the intruder will be detected in the future, then the point $\mathbf{x}$ belongs to what we call the *anti-causal secure zone*. The complete secure zone is thus composed of two regions that may overlap: the causal and the anticausal secure zones. Its complementary $\mathbb{X}$ can be computed by the following contractions

$$\begin{aligned}
\mathbb{X}(t) &= \mathbb{X}(t) \cap \mathbb{O} \\
\mathbb{X}(t) &= \mathbb{X}(t) \cap (\mathbb{X}(t-dt) + dt \cdot \mathbb{F}(\mathbb{X}(t-dt))) \\
\mathbb{X}(t) &= \mathbb{X}(t) \cap \bigcap_i g_{\mathbf{a}_i(t)}^{-1}([d(t), \infty]) \\
\mathbb{X}(t) &= \mathbb{X}(t) \cap (\mathbb{X}(t+dt) - dt \cdot \mathbb{F}(\mathbb{X}(t+dt)))
\end{aligned}$$

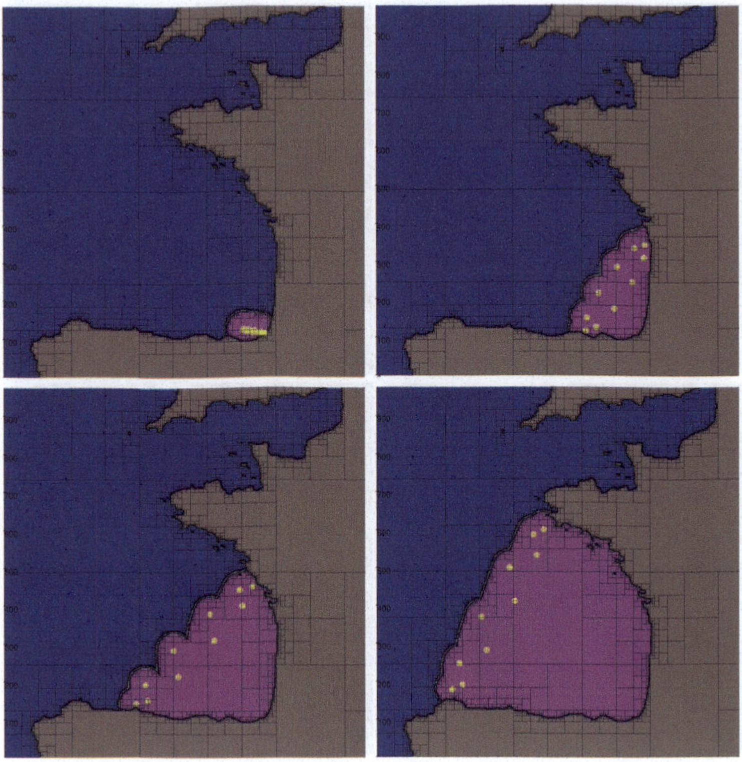

Fig. 7.6 The ellipse moves in order to extend the secure zone (magenta). The position of the robots are represented by yellow points

with the initialization $\mathbb{X}(t) = \mathbb{R}^n$. The contractions should be performed up to the fixed point. This corresponds to a set-membership smoother. As shown in [20] the fixed point is reach in two steps: a forward step followed by a backward step.

7.4.3 Combining with a Barrier Strategy

Each robot follows a reference point. All reference points form a flat ellipsoid which plays the role of a barrier (see Sect. 7.2). The strategy is illustrated by Fig. 7.6 for 10 robots. The set $\mathbb{O}$ corresponds to the blue area (left). On the right, $\mathbb{S}(t)$ is painted green. The observer has been implemented using interval analysis.

7.5 Conclusion

In this paper, we have proposed a set-membership method to compute the secure zone associated to a group of robots. In the secure zone, no undetected intruder can exist. We also proposed a control strategy to make the secure zone as large as possible with a limited number of robots. The principle is to make the group follow a flat ellipse which makes a barrier. Some perspectives to this work are the following

- Find a control strategy based on the shape of the secure zone and also on the shape of the boundary of $\mathbb{O}$. The ellipse strategy is indeed suited to convex shape for $\mathbb{O}$, but cannot deal with more complex shapes. For instance to secure the sea around an archipelago such as UK or Japon cannot be considered with a single barrier.
- Adapt the method to capture an intruder. If we assume that the intruder is captured as soon as it is detected, then we can guess that there exists strong links between securing a zone and the capture problem.
- Consider the escape problem. We are now on the side of the intruder. Escaping from capture by a group of robots or entering inside a zone without being detected can be considered a dual to the problem of securing a zone.

As previously said, we are not totally convinced by our control strategy which is only adapted to convex zones and is not efficient in case of zones with islands (such as archipelago). Since the basic model of Reynolds can been extended in several different ways to incorporate the effects of fear, emotions, leadership change, etc., we would like to take into account the shape of the current secure zone in our control strategy. For instance, we can add a law which motivates each robot to follow a leader (*chain building low*), and another low to follow the boundary of the computed secure zone (*boundary following low*). The strategy could also be take advantage of algorithms taken from operational research such as the solutions of the art gallery problems [21].

Acknowledgements The authors want to thank the following students that where strongly involved in the project: Elouan Autret, Thomas Boulier, Maxime Bouyssou, Philippe Chuzel, Alice Danckaers, David Duverger, Raphael Finkelstein, Sylvain Hunault, Pierre Jacquot, Mael Le Gallic, Eric Mourre, Thiago Oliveira, Benoit Raymond, Khadimoullah Vencatasamy. They made a small video illustrating the project: https://youtu.be/rNcDW6npLfE

References

1. L. Jaulin, *Mobile Robotics*, ISTE editions. (2015)
2. V. Creuze, Robots marins et sous-marins ; perception, modélisation, commande, Techniques de l'ingénieur
3. M. Carreras, P. Ridao, R. García, T. Nicosevici, Vision-based localization of an underwater robot in a structured environment, in *Proceedings of IEEE International Conference on Robotics and Automation ICRA'03* vol. 1 (IEEE, 2003), pp. 971–976
4. S. Bazeille, V. Lebastard, F. Boyer, Different use of electric sense for underwater robots navigation, in *Ocean Engineering and Oceanography*, (Springer, France, 2017)

5. F. Boyer, P.B. Gossiaux, B. Jawad, V. Lebastard, M. Porez, Model for a sensor inspired by electric fish. IEEE Trans. Robot. **28**(2), 492–505 (2012)
6. B. Zerr, G. Mailfert, A. Bertholom, H. Ayreault, Sidescan sonar image processing for auv navigation, in *Oceans 2005-Europe*, Vol. 1 (IEEE, 2005), pp. 124–130
7. A. Caiti, A. Garulli, F. Livide, D. Prattichizzo, Localization of autonomous underwater vehicles by floating acoustic buoys: a set-membership approach. IEEE J. Ocean. Eng. **30**(1), 140–152 (2005)
8. A. Caiti, A. Garulli, F. Livide, D. Prattichizzo, Set-membership acoustic tracking of autonomous underwater vehicles. Acta Acust. United Acust. **5**(88), 648–652 (2002)
9. F.L. Bars, J. Sliwka, O. Reynet, L. Jaulin, State estimation with fleeting data. Automatica **48**(2), 381–387 (2012)
10. L. Jaulin, *Automation for Robotics*, ISTE editions. (2015)
11. H.G. Tanner, A. Jadbabaie, G.J. Pappas, Stable flocking of mobile agents, part I: Fixed topology, in *Proceedings of 42nd IEEE Conference on Decision and Control, 2003*, Vol. 2 (IEEE, 2003) pp. 2010–2015
12. E. Rimon, D.E. Koditschek, Exact robot navigation using artificial potential fields. IEEE Trans. Robot Autom. **8**(5), 501–518 (1992)
13. C. Petres, M.R. Ramirez, F. Plumet, Reactive path planning for autonomous sailboat, in *IEEE International Conference on Advanced Robotics* (2011), pp. 1–6
14. C. Pêtrès, M.-A. Romero-Ramirez, F. Plumet, A potential field approach for reactive navigation of autonomous sailboats. Robot. Auton. Syst. **60**(12), 1520–1527 (2012)
15. J.-P. Aubin, Viability Theory, (Springer Science and Business Media, 2009)
16. M. Kieffer, L. Jaulin, E. Walter, D. Meizel, Nonlinear identification based on unreliable priors and data, with application to robot localization, in *Robustness in Identification and Control*, LNCIS, vol. 245 ed. by A. Garulli, A. Tesi, A. Vicino (Springer, London, 1999), pp. 190–203
17. L. Najman, H. Talbot (ed.), *Mathematical Morphology: From Theory To Applications* (ISTE-Wiley, 2013) (1000)
18. G. Chabert, L. Jaulin, Contractor programming. Artif. Intell. **173**, 1079–1100 (2009)
19. L. Jaulin, B. Desrochers, Introduction to the algebra of separators with application to path planning. Eng. Appl. Artif. Intell. **33**, 141–147 (2014)
20. L. Jaulin, M. Kieffer, I. Braems, E. Walter, Guaranteed nonlinear estimation using constraint propagation on sets. Int. J. Control. **74**(18), 1772–1782 (2001)
21. J. O'Rourke, *Art Gallery Theorems and Algorithms*, (Oxford University Press, 1987)

Chapter 8
Comparison of Kalman and Interval Approaches for the Simultaneous Localization and Mapping of an Underwater Vehicle

Jérémy Nicola and Luc Jaulin

Abstract In this paper we compare the use of a Kalman filter and a Robust State Observer for the localization and mapping of an underwater vehicle using range-only measurements between the vehicle and a set of beacons lying on the seafloor. As expected, we show that the Kalman filter performs great when we have a reasonably good prior information on the location of the vehicle and the beacons. Based on set-membership methods, the Robust State Observer demonstrates an outstanding capacity to provide consistent estimates (where the true solution is in the estimated confidence domain) in the presence of outliers at the cost of a very coarse precision. The source of this lack of precision will be discussed.

8.1 Introduction

Simultaneous Localization and Mapping [1, 2] is the problem of estimating the pose of a vehicle while building a map of its environment. The more precisely the pose is known, the more precise the map will be and reciprocally. In an underwater positioning context, where GPS is not available and SLAM can be seen as a solution for localization. It is possible to compute a crude estimation of the objects positions from the surface, and refine these positions on-the-go as an underwater vehicle navigates between them. It is even possible to avoid completely the prior survey phase, and to go directly for the online calibration, which results in important time savings. SLAM needs a perception of the environment which are based on cameras, sonars and any other sensors. In the context of this paper, the exteroceptive sensor which is used is an acoustic range-only sensor.

J. Nicola
IxBlue, Sainte Anne du Portzic, Brest, France
e-mail: jeremy.nicola@gmail.com

L. Jaulin (✉)
Lab-STICC, UMR CNRS 6285, ENSTA Bretagne, Brest, France
e-mail: luc.jaulin@ensta-bretagne.fr

L. Jaulin et al. (eds.), *Marine Robotics and Applications*, Ocean Engineering & Oceanography 10, https://doi.org/10.1007/978-3-319-70724-2_8

This type of state estimation problem is typically solved with a Kalman filter [3], as it is a well studied, well understood state estimator that has successfully been applied on a number of localization and SLAM related problems [4–6]. While this approach works remarkably well when a good initial estimate is available for the position of the mobile and the beacons, it is known to suffer from poor performances when the system is not well conditioned, when nonlinearities occur and no good initial estimate is known. In these situations, a Kalman filter can produce a poor estimate with a high confidence, and will reject good measurements, being unable to distinguish them from outliers, making it tricky to detect a fault. In this article, we will compare a Kalman filter approach with an other approach based on set-theory [7]. Set-theoretic approaches are interesting because they are probabilistic agnostic and some of them are applicable to nonlinear systems with robustness to outliers. Reference [8] compared the use of a Kalman filter for the localization of a terrestrial vehicle with interval methods (Figs. 8.1 and 8.2).

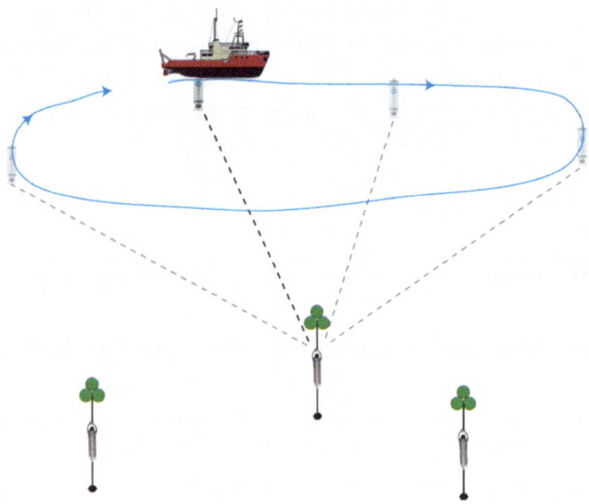

Fig. 8.1 Pre-survey from a surface vessel

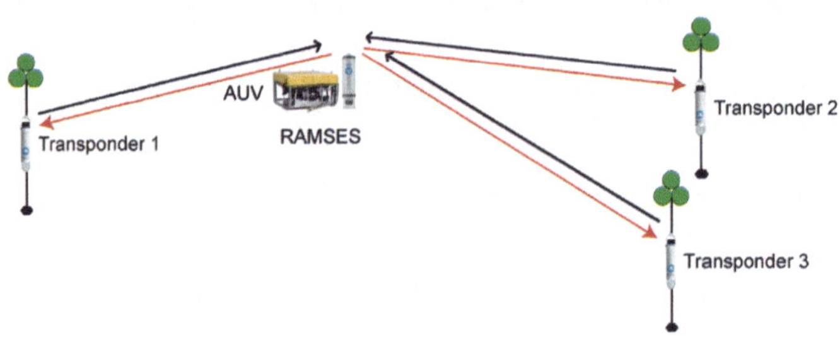

Fig. 8.2 Navigation of an AUV in the calibrated baseline

In this paper, we compare and combine set-membership and probabilistic approaches to solve the SLAM problem for an underwater vehicle. Underwater localization [9, 10] mainly relies on the fact that the time of flight of the sound provides us the distance between the robot and some seamarks. And in this situation, set-membership methods have been shown to be very attractive [11, 12].

The paper is organized as follows. Section 8.2 describes the modelisation of the system and its sensors, with the assumptions on the nature of uncertainties. In Sect. 8.3 we see how our problem can be cast in a probabilistic state estimation problem, and study the use of a Kalman filter to solve it. Section 8.4 briefly recalls the principles behind interval-based state estimation with its theoretical advantages and drawbacks. Section 8.5 compares both methods on a real data set. Section 8.6 discusses the results, and in particular the source for the lack of precision for the interval estimator. Section 8.7 concludes the paper.

8.2 Modeling the Problem

SLAM is a typical state estimation problem [13] that is described by the following state equations [14]:

$$\begin{cases} \mathbf{x}_{k+1} &= \mathbf{f}_k(\mathbf{x}_k) \\ \mathbf{y}_k &= \mathbf{g}_k(\mathbf{x}_k) \end{cases} \tag{8.1}$$

where k is the time, $\mathbf{x} = \left(\mathbf{x}_m, \mathbf{x}_{b_1}, \mathbf{x}_{b_2}, \ldots, \mathbf{x}_{b_N}\right)$ is the state vector, containing the position $\mathbf{x}_m$ of the mobile and the beacons $\mathbf{x}_{b_i}$ and $\mathbf{y}$ is the observation vector. The function $\mathbf{f}$ in the first equation is the evolution function. It models the dynamic part of the problem. The function $\mathbf{g}$ is the observation function that models the exteroceptive measurements the vehicle makes on its environment.

8.2.1 Evolution Equation

In this paper, we work in an underwater environment and we now describe the sensor we use for navigation and SLAM. For its proprioceptive sensors, the vehicle is equipped with a PHINS [15], a fiber optic gyroscope inertial navigation system made by iXBlue [16], a DVL and a pressure sensor. These sensors can be considered as proprioceptive since they are involved in the evolution equations. These proprioceptive sensors can be considered as reliable (i.e., without ouliers) and accurate (Figs. 8.3 and 8.4).

The evolution equation of the mobile can be formulated as:

$$\mathbf{x}_{m_{k+1}} = \mathbf{x}_{m_k} + \mathbf{R}(\phi_k, \theta_k, \psi_k) \cdot \mathbf{u}_k + \boldsymbol{\alpha}_k \tag{8.2}$$

Fig. 8.3 A PHINS 6000 coupled with a DVL

where $\mathbf{R}(\phi, \theta, \psi)$ is the Euler matrix of the vehicle depending on the Euler angle (ϕ, θ, ψ) given by the INS. The vector $\mathbf{u}$ is the speed of the vehicle in the robot frame, and $\boldsymbol{\alpha}$ some noise.

In a typical setup, the sources of uncertainties come from:

- Misalignment or imprecisely known lever arms between the sensors.

Fig. 8.4 The RAMSES acoustic navigation system

- Noise on the collected Euler angles and accelerations.
- Noise on the measured speed.

We consider that the vehicle that runs the experiment is perfectly characterized so that we know precisely the lever arms and the alignments between the sensors. Since the INS gives the attitude of the vehicle with a very high precision, as a first approximation we do not consider the contribution of these noises. The DVL measures the speed of the vehicle by measuring the Doppler shift of a high-frequency acoustic signal projected on the seafloor. Because these measurements are not highly reliable, we consider that they are the main source of uncertainties on the evolution equation of the mobile. Therefore, $\boldsymbol{\alpha}_k$ is modeled as a perturbation whose only contribution is considered to come from the integrated measured speed. Since the beacons are tied to the seafloor, the evolution equation of the ith beacon is simply:

$$\mathbf{x}_{b_{i_{k+1}}} = \mathbf{x}_{b_{i_k}} \tag{8.3}$$

8.2.2 *Observation Equation*

For its exteroceptive sensors the robot uses a classical pressure center which provides the depth of the robot. The vehicle is also equipped with a RAMSES [17], a synthetic baseline acoustic positioning device developed by iXBlue [16] that measures the range between the device and a beacon by measuring the times-of-flight of an acoustic signal sent to, and the acoustic signal answered by the beacon.

Therefore, the observation equation for the seamark observed at instant k is written as:

$$y_k = \sqrt{\left(x_m - x_{b_k}\right)^2 + \left(y_m - y_{b_k}\right)^2 + \left(z_m - z_{b_k}\right)^2} + w_k \qquad (8.4)$$

which simply is a distance equation between the mobile and the beacon, plus some noise w.

The sources of uncertainties in the range measurements come from:

- *Approximate knowledge on the celerity profile*. The celerity profile is a function that maps the depth in the water to the celerity of the sound at this depth. It is used to convert the time of flight of an acoustic signal to a range information between the source and the receiver. It depends on several factors such as the pressure, the temperature, the salinity of the water. Depending on the location and the depth of the emitter and the receiver, the celerity profile can also vary with time.
- *The environment*. No line of sight might be present between the emitter and the receiver, and multipath can be received, introducing outliers in the measurements.
- *Uncertainties on the sensor's internal functioning*. These uncertainties are characterized precisely in laboratory.

Since it is hard to quantify and propagate the uncertainties on the celerity profile and the environment, in this paper we assume that the perturbation on the measurement is additive. This is a gross simplification that works well for the orders of magnitude involved in our experiments. Moreover, due to multiple echos, interferences, unmapped obstacles, etc., outliers may occur.

Exteroceptive outliers is one of the main difficulty met in practice. It that has to be considered in order to build reliable underwater robots.

8.3 Solving with a Kalman Filter

Probabilistic state estimation is the classical approach for state-estimation in robotics [13, 18]. It relies on statistical techniques to integrate imperfect models with imperfect sensing. The most widely used approach for probabilistic state-estimation are the Bayes filters, that offers a methodology for estimating a probability distribution over the state, conditioned on all available data in a recursive manner. The two most widely used Bayesian filters are the Kalman filter and the Particle filter. In this paper, we will focus on Kalman filtering, commonly used in the industry.

8.3.1 Kalman Filtering

The Kalman filter is the best studied approach to implementing a Bayes filter. It represents the knowledge we have about the estimated random variables at time k by their first two statistical moments: the mean $\boldsymbol{\mu}_k$ and the covariance matrix $\boldsymbol{\Gamma}_k$.

8.3.1.1 Prediction Equation

The Kalman filter assumes that the evolution model is linear with additive, normally distributed, white noise expressed by the following equation:

$$\mathbf{x}_{k+1} = \mathbf{A}_k \cdot \mathbf{x}_k + \mathbf{B}_k \cdot \mathbf{u}_k + \boldsymbol{\alpha}_k. \tag{8.5}$$

In our model $\mathbf{A}_k$ is simply the identity matrix at every time k, $\mathbf{B}_k$ is the Euler rotation matrix defined above, and $\mathbf{u}_k$ is the integrated speed. The noise vector is denoted by $\boldsymbol{\alpha}_k$. Since this noise is considered to come from the integrated speed, its covariance matrix is approximated by:

$$\boldsymbol{\Gamma}_{\alpha_k} = \mathbf{R}_k \cdot \boldsymbol{\Gamma}_{v_k} \cdot \mathbf{R}_k^T \tag{8.6}$$

with $\boldsymbol{\Gamma}_{v_k}$ the covariance matrix of the integrated speed expressed in the local frame of the vehicle at time k. Our model is suitable for the use of a Kalman filter, because not only is there no multiplicative noise involved, but the equation is linear with respect to the inputs and the state vector.

The Kalman filter prediction equations are thus

$$\begin{cases} \hat{\mathbf{x}}_{k+1} & = \hat{\mathbf{x}}_k + \mathbf{B}_k \cdot \mathbf{u}_k \\ \boldsymbol{\Gamma}_{k+1} & = \boldsymbol{\Gamma}_k + \boldsymbol{\Gamma}_{\alpha_k} \end{cases} \tag{8.7}$$

When initializing the filter, $\hat{\mathbf{x}}_0$ must be chosen as close as possible to the real position of the vehicle and the beacons. The matrix $\boldsymbol{\Gamma}_0$ must be chosen to reflect the amount of uncertainty on this initial belief. For example, when no good prior information on the localization of a beacon is known, the covariance matrix associated with its position can be set diagonal with very large diagonal elements.

8.3.1.2 Update Equation

For a Kalman filter to apply, the observation equations must be linear in the state vector variables, and subject to additive Gaussian noise. The additive Gaussian noise assumption, while widely used, is a crude simplification of the real noise that is applied to our observations. Indeed, some of those uncertainties, especially the ones on the celerity profile, have no reason to be modeled as Gaussian. The whiteness assumption is also not reasonable: if an object is present between the sensor and a beacon at time k, adding a bias to the measured time of flight, it is very likely that this bias will be present in a neighborhood of k. From Eq. 8.4 it is also clear that the linearity assumption does not hold. There exists mainly two methods to apply a Kalman filter with nonlinear models:

- Extended Kalman Filter: this approach consists of linearizing the equations by Taylor series expansion around of the current estimated mean to be able to apply the Kalman filter update equations.

- Unscented Kalman Filter [19]: this approach propagates a finite set of sample points, called sigma points, around the mean through the nonlinear function, from which the posterior mean and covariance matrix of the estimate are computed. This flavor captures more accurately the true mean and covariance of the estimate.

However, even if the observation noise is really Gaussian, the probability density function of the estimate is very unlikely Gaussian due to the nonlinearities. Therefore, the two first statistical moments might not contain enough information to correctly represent the true estimated distribution.

8.3.1.3 Dealing with Outliers

A criteria can be defined on the innovation of a specific observation to decide if we should accept or reject it. With $\tilde{\mathbf{y}}_k$ the innovation at time k, defined by the difference between the predicted and the actual observation and, $\mathbf{S}_k$ the covariance of the innovation, and a confidence threshold η, we consider the condition

$$\tilde{\mathbf{y}}_k \cdot \mathbf{S}_k^{-1} \cdot \tilde{\mathbf{y}}_k^T < \alpha(\eta). \tag{8.8}$$

If this condition is not satisfied, that is, if the observation is too distant from the predicted observation, the measurement $\mathbf{y}_k$ is rejected. The quantity $\alpha(\eta)$ gives the radius of the confidence ellipsoid given a confidence threshold η. This criteria is also found under the name of Mahalannobis distance in the literature. It works remarkably well when the estimate is "close" to the real solution. But when it is not the case, this criteria will eventually discard good observations, because the estimate has not converged to the real solution yet. This is especially the case when no good initialization is available for the filter. Empirical strategies must then be put in place to allow for some of these measurements to be integrated anyway. Note that the existence of outliers is not rare in an underwater acoutic system due to multiple echos, interferences, etc.

8.4 Solving with Sets

Set-membership methods propose a non-stochastic approach to treat uncertainties, by making the assumption that the noises are unknown but bounded. The estimated variables are enclosed in sets. Different representations for sets are available: zonotopes [20], ellipsoids [21, 22], intervals [23, 24], subpavings [25, 26]. Such representations have several advantages:

- Since only the support of the density distribution function for the uncertainties are considered, they are compatible with an infinite number of distribution functions.
- For the case of interval and subpaving representations, no linearization is required since these representation can be propagated through any nonlinear function.

- For the case of intervals and subpavings, there exist a scheme for dealing with outliers based on the q-relaxed intersection [27].

Since in this paper we face the problem of unknown noises and nonlinearities, we retained the interval and subpaving representations. Reference [23] presents a state-estimator based on interval analysis which alternates prediction and correction steps the same way as a Kalman filter. Reference [28] compares this approach with an approach based on constraint propagation. Both methods should be combined and made robust with respect to outliers as proposed in [7], which is the chosen approach for this paper.

8.4.1 Principle

It is assumed that the initial state $\mathbf{x}_0$ belongs to a known set $\mathbb{X}_0$, and that $\mathbf{y}(k)$ belongs to some sets $\mathbb{Y}(k)$ (which are intervals). The set $\mathbb{X}(k)$ containing all the feasible state vectors at time k can be computed recursively [29] by the relation

$$\mathbb{X}(k+1) = \mathbf{f}_k \left(\mathbb{X}(k) \cap \mathbf{g}^{-1}(\mathbb{Y}(k))\right).$$

W can also consider the following equation

$$\mathbb{X}(k+1) = \mathbf{f}_k(\mathbb{X}_k) \cap \bigcap_{i \in \{0,\ldots,\ell\}} \mathbf{f}_k^i \circ \mathbf{g}_k^{-1}(\mathbb{Y}(k-i)) \tag{8.9}$$

with

$$\mathbf{f}_k^i = \mathbf{f}_k \circ \mathbf{f}_{k-1} \circ \cdots \circ \mathbf{f}_{k-i}. \tag{8.10}$$

This equation defines the set of for all state vectors that are consistent with all data inside a time window of length ℓ.

8.4.2 Dealing with Outliers

It happens that some $\mathbf{y}(k)$, the actual value of the observed quantity at time k, do not belong to their corresponding set $\mathbb{Y}(k)$. $\mathbf{y}(k)$ is said to be an inlier if $\mathbf{y}(k) \in \mathbb{Y}_k$, and an outlier otherwise. With the assumption that there are at most q outliers among the last l observations, the set given by the$q$-relaxed intersection $\overset{q}{\bigcap}_{i \in \{0,\ldots,l\}} \mathbf{f}_k^i \circ \mathbf{g}_k^{-1}(\mathbb{Y}(k-i))$ is guaranteed to contain $\mathbf{x}(k)$. The q-relaxed intersection $\overset{q}{\cap}$ is a set operation that computes the classical intersection of m sets except q of them (Fig. 8.5).

Computing the q-relaxed intersection of n intervals is solved in $O(n.log(n))$. However, when q is not fixed, the complexity of computing the q-relaxed intersection

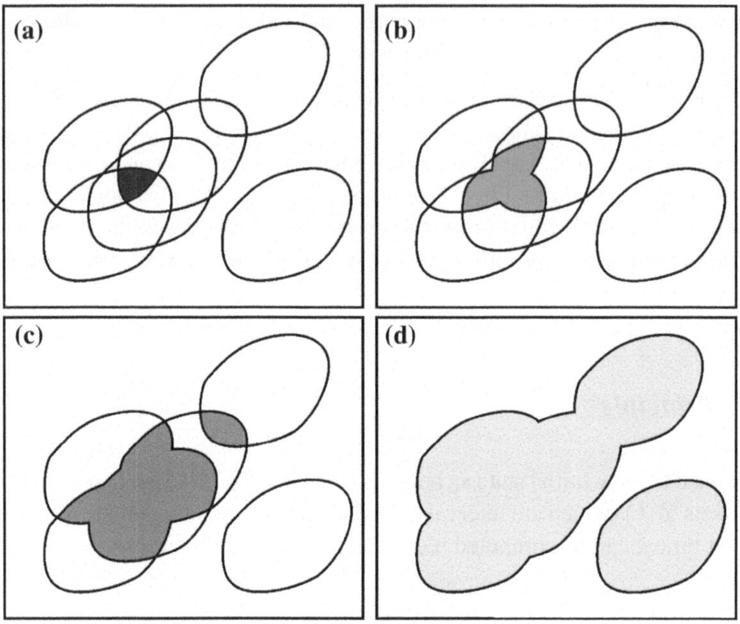

Fig. 8.5 Q-relaxed intersection of 6 sets with respectively **a** $q = 2$, **b** $q = 3$, **c** $q = 4$, **d** $q = 5$

grows exponentially with the dimension of the boxes. Algorithms that compute an overestimation of the true q-intersection have to be used. For a complete review of the q-intersection algorithms the reader is referred to [27, 30].

8.4.3 Robust State Estimator

Using the tools defined in Sects. 8.4.1 and 8.4.2, we propose the Robust State Estimator (RSO) [7]

$$\text{RSO:} \begin{cases} \mathbb{X}(k) = & \mathbf{f}_k^0(\mathbb{X}_0) \text{ if } k < m(initialization) \\ \mathbb{X}(k) = & \mathbf{f}_k^{k-m}(\mathbb{X}(k-m)) \cap \\ & \bigcap_{i\in\{0,\ldots,\ell\}}^{\{q\}} \mathbf{f}_k^i \circ \mathbf{g}_k^{-1}(\mathbb{Y}(k-i)) \text{ if } k \geq m \end{cases} \tag{8.11}$$

where q is the number of outliers that are allowed inside the time window of length ℓ. This estimator is proved to be robust with respect to outliers.

8.5 Comparison

In this section, we compare the two approaches on a dataset acquired near La Ciotat, France in February 2014. A ship (the *vehicle*) is equipped with a PHINS [15], a RAMSES [17], a DVL, a GPS and an acoustic modem to measure distances between existing beacons. It performs a survey of a zone where 4 acoustic beacons lie on the seafloor. The position of these 4 seamarks are precisely known. The trajectory of the ship and the position of the seamarks are depicted in Fig. 8.6. This trajectory is obtained by a fusion of the GPS with the INS sensors. It can be considered as our ground truth. The vehicle (the ship) plays here the role of an underwater robot. Since it is always at the surface, a pressure sensor is not needed. We assume here that the depth is measured as zero with an error of ± 1 m.

8.5.1 Methodology

We will first run a test-case where the vehicle position is assumed to be perfectly known since the start of the mission, and the beacons positions are known up to a bias in the order of magnitude of 30 m in the X/Y plane. We will then run a test-case where the vehicule position is still assumed to be perfectly known, but the seamarks X/Y coordinates are supposedly totally unknown. In this case, the Kalman filter for the seamarks is initialized on the initial position of the vehicle with a very large

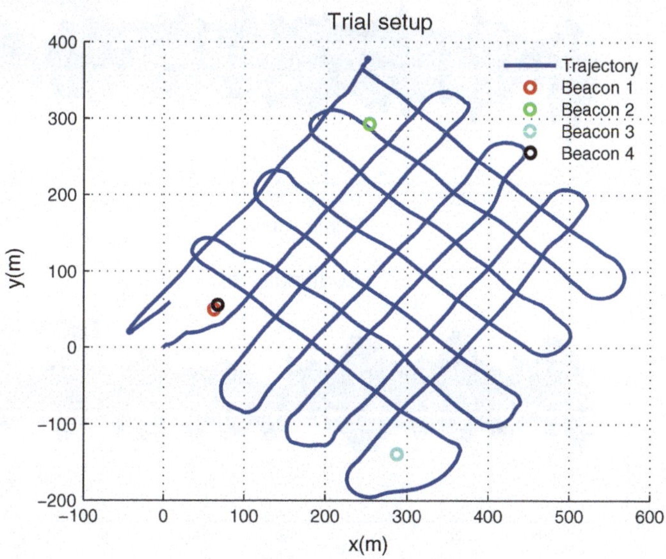

Fig. 8.6 Illustration of the experiment used for the comparison

covariance matrix. In both scenarios, the altitude of the beacons is known with a precision of about one meter. We will compare the two approaches for the estimation of the vehicle's and the seamark's positions.

Consistency. An estimation is said to be *consistent* if the true value of the estimated quantity is contained in the confidence domain provided by the filter. For the Kalman filter, we will say that an estimated position is consistent if the true position is inside its 99% confidence ellipse. For the RSO, the estimation is consistent if it is contained in its subpaving. We will compare the rate of consistency for both filters during all the mission.

Error. We define the estimation error as the Euclidian distance that separates it from the true value. For the Kalman filter, it will be the distance between its mean and the true position, for the RSO it will be the distance between the center of mass of the subpaving and the true position.

8.5.2 SLAM with Some Initial Knowledge

Figure 8.7 displays the confidence domains for the error on the x (Easting) and y (Northing) positions of the vehicle in the local frame for the Kalman filter and the

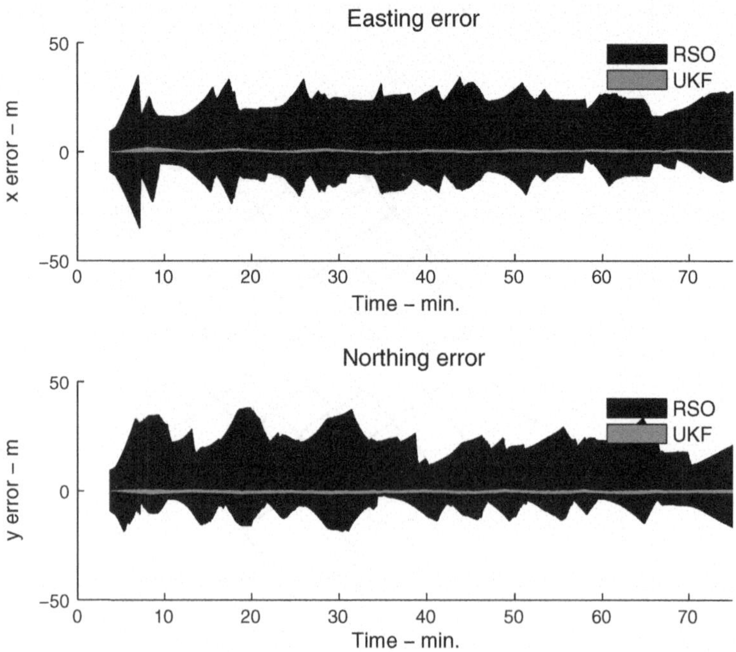

Fig. 8.7 Confidence for the Easting and the Northing of the trajectory when the base is biased

Table 8.1 Precision for the vehicle's localization when a small bias is added to the beacon's positions

Vehicle's precision	UKF	RSO
Max. Error—m	1.15	71.48
Final Error—m	0.469	9.57
Consistency—%	100	100

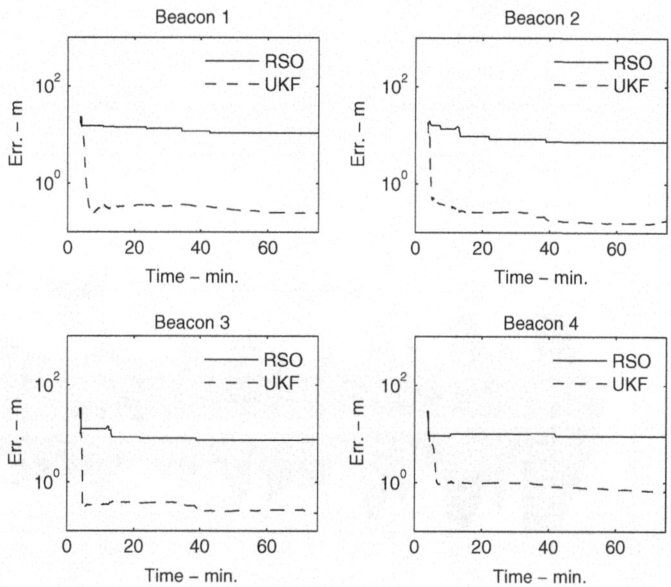

Fig. 8.8 Estimation errors of the beacon's positions when their initial positions are biased

RSO. The Kalman filter gives an estimate with an decimetric error, while the RSO gives a decametric errors. Both approaches contain the real position during all the mission (Table 8.1).

Regarding the estimated positions of the beacons shown in Fig. 8.8, the Kalman filter converges in less than 10 min to a decimetric precision, while the RSO's precision quickly reaches a decametric precision and doesn't improve after that. Both approaches provide consistent estimations: the corresponding sets contain the true positions of the beacons during all the mission (Table 8.2).

8.5.3 SLAM Without Initial Knowledge

Figure 8.9 shows that the confidence on the error of the vehicle's position is higher than for the previous scenario. However, both filters are consistent during the whole

Table 8.2 Precision for the beacons localization when a small bias is added to their initial positions

Beacons precision		UKF	RSO
Beacon 1	Final Error—m	0.246	10.99
	Initial bias—m	28.7	
	Consistency—%	100	100
Beacon 2	Final Error—m	0.180	7.45
	Initial bias—m	22.8	
	Consistency—%	100	100
Beacon 3	Final Error—m	0.221	7.56
	Initial bias—m	35.1	
	Consistency—%	100	100
Beacon 4	Final Error—m	0.657	9.05
	Initial bias—m	30.1	
	Consistency—%	40.63	100

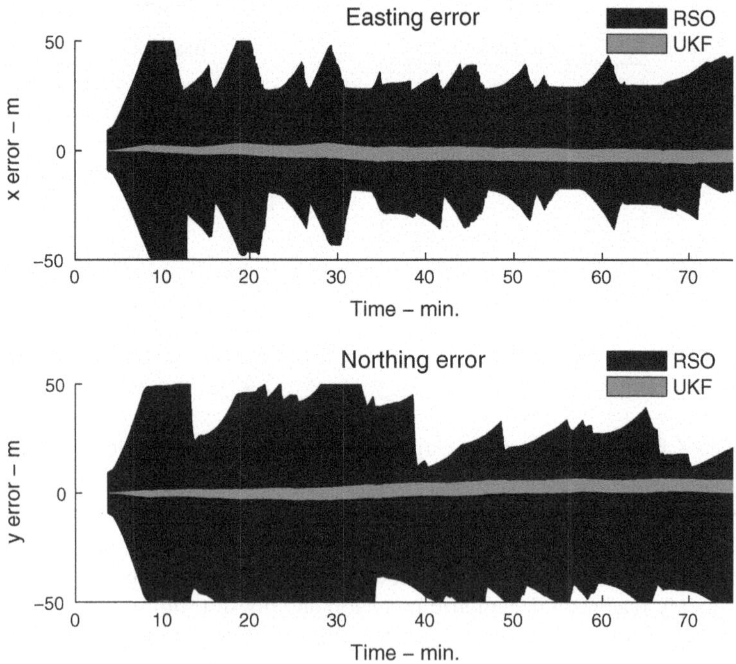

Fig. 8.9 Confidence the Easting and the Northing of the vehicle when the base is unknown

mission: their 99% confidence domains for the vehicle's positions contain the true position. Regarding the estimation of the beacon's positions, the Kalman filter is able to locate the second beacon with a final error of about 4 meters, while it converges to wrong positions for the other beacons, whose 99% confidence domains do not

Table 8.3 Precision for the vehicle's localization when the beacon's positions are unknown

Vehicle's precision	UKF	RSO
Max. Error—m	4.76	71.48
Final Error—m	3.73	19.0
Consistency—%	83.9	100

contain the true positions. The RSO on the other hand, is able to locate the beacons with a final error of about 10 m for all the beacons, and the true beacon's positions are contained in its estimate (Table 8.3).

8.6 Discussion on the Pessimism of the RSO

Section 8.5 has shown that the RSO's precision, despite having a 100% consistency, is quite unsatisfying. One might argue that its high consistency comes from its high level of pessimism. In this section, we describe two factors that have been identified as a unnecessary source of pessimism (Fig. 8.10).

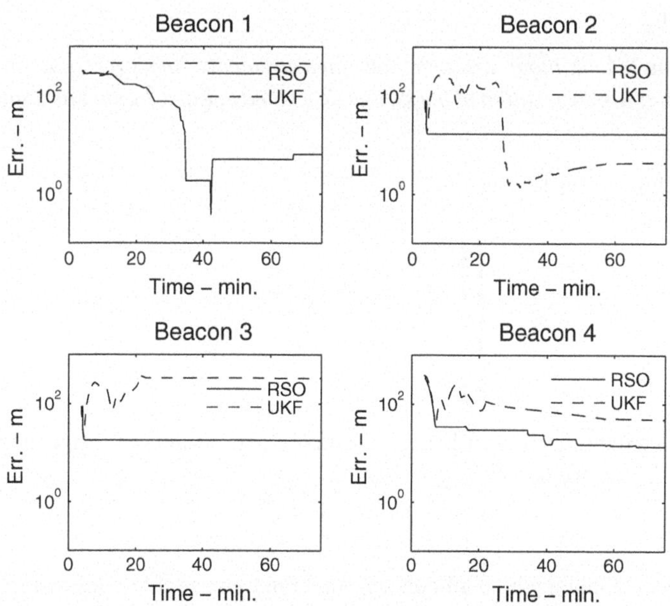

Fig. 8.10 Precision for the beacons when the base is unknown

8.6.1 Wrapping Effect

Definition 8.1 A transformation T is said to be box-conservative if the image of an axis-aligned box through T is also an axis-aligned box.

Remark 8.1 The evolution model for the vehicle involves a rotation that maps the speed measured in the local vehicle's frame to the global frame. A rotation is generally not box-conservative, and this rotation adds pessimism known as wrapping-effect [31].

Since the evolution model is linear, intervals are not the best representation for sets. For this kind of models, other methods such as zonotopes or ellipsoids are a more attractive representation. To reduce the wrapping induced by the rotation, a combination of linear and interval methods could be used.

8.6.2 Non-compensation of Noises

Assume the robot moves in a fixed direction. We have $\forall k, \mathbf{R}_k = \mathbf{R}, \boldsymbol{\Gamma}_{\alpha_k} = \boldsymbol{\Gamma}_{\alpha}$. At the beginning the robot knows its exact position, so $\boldsymbol{\Gamma}_0 = 0$. If the robot does not make any observation, we have the following properties. a unnecessary source of pessimism.

Proposition 8.1 *Without exteroceptive measurements, the precision given by a Kalman filter for the position of the system described in Eq. 8.2 grows as a square-root of the time.*

Proof Following Eq. 8.7, we have

$$\begin{cases} \boldsymbol{\Gamma}_1 & = \boldsymbol{\Gamma}_\alpha \\ \boldsymbol{\Gamma}_2 & = \boldsymbol{\Gamma}_1 + \boldsymbol{\Gamma}_\alpha = 2 \cdot \boldsymbol{\Gamma}_\alpha \\ \vdots & \\ \boldsymbol{\Gamma}_k & = k \cdot \boldsymbol{\Gamma}_\alpha \end{cases} . \tag{8.12}$$

Given a fixed time-step dt between instant k and instant $k+1$, the elapsed time is $t = k \cdot dt$, and we have

$$\boldsymbol{\Gamma}_k = \frac{t}{dt} \cdot \boldsymbol{\Gamma}_\alpha. \tag{8.13}$$

A confidence domain with confidence η for a random variable x following a normal distribution with mean μ and variance σ^2 is described by the following inequality

$$(x - \mu)^2 \leq \alpha^2(\eta) \cdot \sigma^2 \tag{8.14}$$

Table 8.4 Precision for the beacons localization when the initial location of the vehicle is unknown

Beacons precision		UKF	RSO
Beacon 1	Final Error—m	458.2	6.40
	Initial bias—m	319.8	
	Consistency—%	11.6	100
Beacon 2	Final Error—m	4.29	16.85
	Initial bias—m	80.4	
	Consistency—%	68.9	100
Beacon 3	Final Error—m	332.7	17.83
	Initial bias—m	87.1	
	Consistency—%	24.6	100
Beacon 4	Final Error—m	48.7	13.52
	Initial bias—m	387.6	
	Consistency—%	17.5	100

with $\alpha(\eta)$ a function that gives the confidence threshold α for a confidence η. Consequently, for any of the mobile's position component x_i, we have

$$\sqrt{\frac{\left(x_i - \hat{x}_i\right)^2}{\alpha^2(\eta) \cdot \Gamma_{\alpha_{ii}}} \cdot dt} \leq \sqrt{t} \tag{8.15}$$

which concludes the proof (Table 8.4).

Proposition 8.2 *Without exteroceptive measurements, the precision given by the RSO for the position of the system described in Eq. 8.2 grows linearly with time.*

Proof The perturbation on the integrated speed in Eq. 8.1 is supposed to belong to some interval $[\mathbf{w}_\alpha]$.

From we have

$$\begin{cases} \left[\boldsymbol{x}_{m_1}\right] &= \boldsymbol{x}_{m_0} + \mathbf{R}\cdot\mathbf{u} + [\mathbf{w}_\alpha] \\ \left[\boldsymbol{x}_{m_2}\right] &= \left[\boldsymbol{x}_{m_1}\right] = \boldsymbol{x}_{m_0} + 2\cdot\mathbf{R}\cdot\mathbf{u} + 2\cdot[\mathbf{w}_\alpha] \\ \vdots & \\ \left[\boldsymbol{x}_{m_2}\right] &= \boldsymbol{x}_{m_0} + k\cdot\mathbf{R}\cdot\mathbf{u} + k\cdot[\mathbf{w}_\alpha] \end{cases} \tag{8.16}$$

Given a fixed time-step dt between instant k and instant $k+1$, the elapsed time is $t = k \cdot dt$, and with $w([x])$ the width of an interval $[x]$, we have

$$w\left(\left[x_{m_{k_i}}\right]\right) = \frac{t}{dt} \cdot w\left(\left[w_{\alpha_i}\right]\right)$$

which concludes the proof.

Propositions 8.1 and 8.2 shows that the Kalman filter is much more precise when integrating the proprioceptive measurements than its interval counterpart. By taking into account that the proprioceptive noises are Gaussian, it is able to provide a precision that grows as a square root of the time, whereas the RSO provides an estimate whose precision grows linearly with time.

8.7 Conclusion

In this paper, we compared the use of a Kalman filter against an interval filter for the SLAM problem of a vehicle (here a ship) on a real data set. As expected, the Kalman filter gives estimates that are of high precision when we have a reasonable prior knowledge on the beacons positions, and might converge towards wrong solutions when no such knowledge is available. The RSO, on the other hand, is consistent during all the mission with or without good prior knowledge, but its precision is quite poor. To improve the precision of the Kalman filter, we could use an approach similar to [32], where the initial state is first estimated with an interval method, and then a Kalman filter is initialized from this estimate. An other approach would be to overcome the sources of pessimism described in Sect. 8.6. Since the evolution model is linear, the wrapping effect could be reduced by combining an interval approach with a linear approach such as ellipsoidal methods [22, 29, 33]. An other way of drastically reducing the pessimism of the RSO would be to make it able to integrate the uncertainties as a square-root of time, which would make the propagation-retropropagation described in Eq. 8.10 much more precise.

References

1. R.C. Smith, P. Cheeseman, On the representation and estimation of spatial uncertainly. Int. J. Rob. Res. **5**(4), 56–68 (1986)
2. J. Leonard, H. Durrant-Whyte, Simultaneous map building and localization for an autonomous mobile robot, in *Proceedings of the IEEE/RSJ International Conference on Intelligent Robots and Systems, IROS 1991, November 3–5, 1991* (Osaka, Japan, 1991), pp. 1442–1447
3. R.E. Kalman, A new approach to linear filtering and prediction problems. Trans. ASME J. Basic Eng. **82**(Series D), 35–45 (1960)
4. G.A. Kantor, S. Singh, Preliminary results in range-only localization and mapping, in *Proceedings of the IEEE Conference on Robotics and Automation (ICRA'02)*, vol. 2 (May 2002), pp. 1818–1823
5. E. Olson, J. Leonard, S. Teller, Robust range-only beacon localization. IEEE J. Ocean. Eng. **31**(4), 949–958 (2006)
6. J.J. Leonard, H.F. Durrant-Whyte, Mobile robot localization by tracking geometric beacons. IEEE Trans. Robot. Autom. **7**(3), 376–382 (1991)
7. L. Jaulin, Set-membership localization with probabilistic errors. Robot. Auton. Syst. **59**(6), 489–495 (2011)
8. A. Gning, P. Bonnifait, Dynamic vehicle localization using constraints propagation techniques on intervals a comparison with kalman filtering, in *Proceedings of the 2005 IEEE International*

Conference on Robotics and Automation, ICRA 2005, April 18–22, 2005 (Barcelona, Spain, 2005), pp. 4144–4149
9. B. Zerr, G. Mailfert, A. Bertholom, H. Ayreault, Sidescan sonar image processing for auv navigation, in *Oceans 2005-Europe*, (vol. 1), (IEEE, 2005), pp. 124–130
10. V. Creuze, Robots marins et sous-marins; perception, modélisation, commande, *Techniques de l'ingénieur* (2014)
11. A. Caiti, A. Garulli, F. Livide, D. Prattichizzo, Localization of autonomous underwater vehicles by floating acoustic buoys: a set-membership approach. IEEE J. Ocean. Eng. **30**(1), 140–152 (2005)
12. A. Caiti, A. Garulli, F. Livide, D. Prattichizzo, Set-membership acoustic tracking of autonomous underwater vehicles. Acta Acust. united Acust. **5**(88), 648–652 (2002)
13. L. Jaulin, *Automation for Robotics*. ISTE editions (2015)
14. A. Caiti, A. Garulli, F. Livide, D. Prattichizzo, *Mobile Robotics*. ISTE editions (2015)
15. PHINS product page, https://www.ixblue.com/products/phins. Accessed 07 Dec 2015
16. iXBlue website, http://ixblue.com. Accessed 07 Dec 2015
17. RAMSES product page, https://www.ixblue.com/products/ramses. Accessed 07 Dec 2015
18. S. Thrun, W. Bugard, D. Fox, *Probab. Robot.* (MIT Press, Cambridge, M.A., 2005)
19. S.J. Julier, J.K. Uhlmann, A new extension of the kalman filter to nonlinear systems (1997), pp. 182–193
20. M. Maiga, N. Ramdani, L. Travé-Massuyès, C. Combastel, A csp versus a zonotope-based method for solving guard set intersection in nonlinear hybrid reachability. Math. Comput. Sci. **8**(3–4), 407–423 (2014)
21. B.T. Polyak, S.A. Nazin, C. Durieu, E. Walter, Ellipsoidal parameter or state estimation under model uncertainty. Automatica **40**(7), 1171–1179 (2004)
22. F. Chernousko, Ellipsoidal state estimation for dynamical systems. Nonlinear Anal. Theory Methods Appl. **63**(5–7), 872–879 (2005)
23. M. Kieffer, L. Jaulin, E. Walter, D. Meizel, Guaranteed mobile robot tracking using interval analysis, in *Proceedings of the MISC'99 Workshop on Applications of Interval Analysis to Systems and Control* (Girona, Spain, 1999), pp. 347–359
24. O. Lévêque, L. Jaulin, D. Meizel, E. Walter, Vehicule localization from inaccurate telemetric data: a set inversion approach, in *Proceedings of 5th IFAC Symposium on Robot Control SY.RO.CO.'97* (vol. 1) (Nantes, France, 1997), pp. 179–186
25. V. Drevelle, L. Jaulin, B. Zerr, Guaranteed characterization of the explored space of a mobile robot by using subpavings, in *Proceedings Symposium Nonlinear Control Systems (NOLCOS'13)* (Toulouse, 2013)
26. L. Jaulin, Range-only SLAM with occupancy maps; A set-membership approach. IEEE Trans. Robot. **27**(5), 1004–1010 (2011)
27. C. Carbonnel, G. Trombettoni, P. Vismara, G. Chabert, Q-intersection algorithms for constraint-based robust parameter estimation, in *Proceedings of AAAI*, ed. C. o. A. I (AAAI Press, Québec City, Canada, 2014), pp. 2630–2636
28. B. Vincke, A. Lambert, Experimental comparison of bounded-error state estimation and constraints propagation, in *IEEE International Conference on Robotics and Automation, ICRA 2011, 9–13 May 2011* (Shanghai, China, 2011)
29. D.P. Bertsekas, I.B. Rhodes, Recursive state estimation for a set-membership description of uncertainty. IEEE Trans. Autom. Control **16**(2), 117–128 (1971)
30. L. Jaulin, E. Walter, Guaranteed robust nonlinear minimax estimation. IEEE Trans. Autom. Control **47**(11), 1857–1864 (2002)
31. N.S. Nedialkov, K.R. Jackson, A new perspective on the wrapping effect in interval methods for initial value problems for ordinary differential equations, in *Perspectives on Enclosure Methods* (Springer-Verlag, 1999), pp. 219–264
32. D. Bouvet, G. Garcia, Guaranteed 3-d mobile robot localization using an odometer, an automatic theodolite and indistinguishable landmarks, in *Robotics and Automation, 2001. Proceedings 2001 ICRA. IEEE International Conference on*, vol. 4 (2001), pp. 3612–3617

33. C. Durieu, B. Polyak, E. Walter, Ellipsoidal state outer-bounding for MIMO systems via analytical techniques, in *Proceedings of the IMACS—IEEE–SMC CESA'96 Symposium on Modelling and Simulation*, vol. 2 (Lille, France, 1996), pp. 843–848

Chapter 9
Evolutionary Dynamic Reconfiguration of AUVs for Underwater Maintenance

O. Chocron, E. P. Vega and M. Benbouzid

Abstract In this chapter, the in-mission optimization of the propulsion and control systems of AUVs is developed. This optimization is aimed at obtaining parameters to adapt a dynamically reconfigurable propulsion to underwater infrastrtuctures monitoring missions. The hydrodynamic model of the AUVs is examined. Additionally, a four-thrusters fixed propulsion system is proposed and modelled. A control method using the hydrodynamic model is developed and adapted to our AUV topology. The optimization is used to find suitable vectorial-thrusters orientations and control parameters in order to execute given robotic tasks, speeding up the convergence and minimizing the energy consumption. This is done using a genetic algorithm, which is a stochastic optimization method used here for task-based design. The optimization results are used for readily adapt the vectorial propulsion system to different tasks specificatins, making the AUV dynamically reconfigurable.

9.1 Introduction

When designing a robot to perform precise missions, said sub-systems and parameters must be taken into account in order to create a robot adapted to the user needs. The designer, faced with a blank sheet of paper at the beginning of the project, needs to imagine the missions the robot will perform and come up with an appropriate system. This system will be a carefully chosen combination of the design elements such as hull shape, propulsion system or control.

O. Chocron (✉) · E. P. Vega
Ecole Nationale d'IngÃl'nieurs de Brest (ENIB), FRE CNRS 3744 IRDL, TechnopoÌĆle Brest Iroise, 29238 Brest Cedex 03, France
e-mail: chocron@enib.fr

E. P. Vega
e-mail: vega@enib.fr

M. Benbouzid
University of Brest, FRE CNRS 3744 IRDL, 29238 Brest Cedex 03, France
e-mail: Mohamed.Benbouzid@univ-brest.fr

L. Jaulin et al. (eds.), *Marine Robotics and Applications*, Ocean Engineering & Oceanography 10, https://doi.org/10.1007/978-3-319-70724-2_9

Given the large amount of design elements, and their heterogeneity, finding a suitable configuration is a challenging task. If additionally, we are looking for a highly performing configuration, the design process becomes extraordinarily complex and no known systematic method exists.

To overcome such a complexity, the designer can choose from the following strategies:

- Use a descriptive approach, which is the classical "human" approach. The designer uses previous information and novel ideas in order to solve the problem. The ideas are transformed iteratively in the mind of the designer until a satisfactory solution is created. Since this process occurs in a semi-empirical way, the final design is as good as the experience and capacity of the designer.
- Use a digital approach, which consists on giving to a computer the tasks of creating a system or analyzing a previously designed one. This approach provides several advantages. It can be used to automatically find the design elements of a system that optimizes a certain characteristic. Computer aided design (CAD) techniques, take precise instructions about the features to optimize and, using an iterative process, find the most fitted solution(s).

Taking into account the number of parameters and sub-systems to determine before properly designing a robot propulsion, it is clear that such a difficult task would greatly benefit from a computer-based optimization design method. Namely, the design problem discussed in this research work, can be formulated as a global optimization problem.

9.2 Underwater Robots Propulsion Design

9.2.1 Propulsive Topologies

The degree of actuation, the amount of degrees of freedom actuated in the AUV, defines what kind of movements can be generated by the robot propulsive system (propulsive topology). Evidently, the more an underwater robot is actuated, the more complex tasks it will be able of perform.

At the design stage, the designer needs to imagine the type of tasks that the AUV will perform in order to determine the needed degree of actuation of the vehicle. This task, simple at first sight, is decisive not only for the final capabilities of the robot but also for its overall performance. Indeed, a poorly chosen propulsive configuration can lead to an AUV incapable of performing certain tasks or, at the least, to an under-performant vehicle.

To make sure an AUV is maneuverable enough to perform a great amount of tasks, an immediate approach would be actuate all six degrees of freedom of the Cartesian space, making it *fully actuated*. However, as stated in the introduction of this thesis, another desired characteristic of an AUV is a high degree of autonomy. An AUV actuated in all of its degrees of freedom will certainly need many actuators, which would make the vehicle power-consuming. This could cause a rapid consumption of the, necessarily limited, energetic resources of the robot.

In the process of creating an AUV, at some point the designer must face this dichotomy between actuation capabilities and autonomy (power consumption). This is a key point in the development of the robot, since the solution to this dilemma will be fundamental for the rest of the process.

Several kind of topologies can be found in the literature. This shows the importance of this design domain, as well as the lack of a general solution capable of satisfying most application cases. We can divide these robots in two groups: fully actuated and underactuated robots.

9.2.1.1 Fully Actuated Robots

This type of underwater robots have six or more actuators actuating the six degrees of freedom of the robot. This property allows these robots to perform a great variety of tasks, since with their propulsive resources they can move in the space with complete freedom (or holonomy). This freedom comes, however, at the cost of consuming great amounts of energy.

Despite their outstanding maneuverability, there are relatively few fully actuated AUV. This could be because of their power consumption or due to the fact that their potential agility is not yet matched with reliable control and navigation systems. Another reason for the reduced number of fully actuated AUVs can be the lack of appropriate missions. Indeed, for a great number of tasks nowadays there is not a valid reason to actively control all of the degrees of freedom of an AUV. This is due to the fact that AUV operations are mostly meant to be planar and at a certain depth, without many changes in orientation.

The importance of this type of vehicles; however, can not be neglected. A testimony of this, are the research efforts made in order to correctly control this AUV category. Some of these works mostly take a general approach at the control of these vehicles, aiming their developments to a conceptual level and not focusing on the robot particularities nor on a direct application [1–3], while others tackle the control of an specific vehicle [4].

Industrial AUV tend to focus on survey-style tasks [5–7], in which maneuverability is not as important as autonomy and velocity. In such an scenario, fully actuated AUVs are not yet sufficiently attractive. However, for academic proposes, fully actuated AUVs are very appealing. Indeed, due to their maneuverability, these vehicles can be used as a test bench in order to develop new control and localization methods [8–10]. Evidently, given the maneuverability of these robots and thanks to new advances in control, these underwater vehicles are receiving more attention from companies willing to invest in this kind of technology [11].

9.2.1.2 Under Actuated Robots

The second type of robots reviewed in this section are called *underactuated*. The main characteristic of thee AUV is that they are not able to fully control all of their six degrees of freedom.

The amount of underactuated AUVs found in the literature is greater than the amount of their fully actuated counterparts. This could be due to the reduced power consumption (less actuators means less consumed energy) or to the fact that most of the tasks for these robots only need the control over a few degrees of freedom.

Many examples of underactuated AUVs are available in the literature. Historically, the most common type were the torpedo-shaped underwater robots [12]. These vehicles, designed to travel great distances in order to perform survey-style missions, were the first AUVs to find their way out of research facilities and get adopted to commercial use [13, 14]. The range of applications of torpedo-shaped AUVs is quite large, they are used to collect data [15], to map the sea floor [16] and to inspect pipelines [17]. Motivated by the need of autonomy and to travel great distances, a type of torpedo-shaped AUV called *glider* has been developed in the last decades [18, 19]. The main advantages of these vehicles are the long range mission capabilities with low energy consumption.

After the first wave of torpedo-shaped AUVs and thanks to advancements in control, localization and path planing, AUVs applications have mutated to other areas of interest. Along with the change of mission, the structure of these AUVs have mutated to adapt to new missions. Nowadays, we can find (mostly underactuated) AUVs acting as diving companions [20], cleaning ship hulls [21] and reaching their application scope to intervention tasks [22]. Even if they seem complex, these tasks do not demand the control of all degrees of freedom of the robot.

Control over this type of vehicle is more challenging than over fully actuated robots. Indeed, for underactuated AUVs, not every trajectory is reachable, due to their propulsive limitations. Great efforts have been made in order to overcome this drawback, in the literature we can find advanced and creative control techniques [23–26] trying to solve this problem.

9.2.2 Propulsion Technology

Equally important as the propulsive technology, is the way in which the propulsive forces are created. In this section, different propulsive means will be discussed, ranging from classic technologies, such as control and control surface, to the innovative biomimetic approach.

9.2.2.1 Fixed Propulsion

The first type of propulsion technology analyzed in this section is the fixed one. These actuators are able to generate the propulsive force along a fixed direction with regard to the robot body. Usually, in order to properly control the degrees of freedom required by a mission, many fixed actuators are needed.

The number, orientation and position of these actuators on the hull of the AUV depend on the type of mission performed by the robot. In virtue of this, no general rules exist to define said design parameters. This step of the design process relies heavily on the previous experience of the designer.

Several ways of creating a fixed propulsive force exist [27]. The two most common being thrusters and water-jet actuators.

Thrusters are nowadays the most used technology, their relatively low cost and the fact that they have been developed for decades, give these actuators great advantages in terms of technological maturity. Given the popularity of these actuators, we can find them mounted o AUVs of very diverse types [28–30].

Also due to their popularity, several mathematical models explaining the behavior of these devices exist [31–33], which encourages their use even more. A sign of popularity of these actuators are the diverse studies we can find in the literature. These go from researching the hydrodynamic interaction between the thruster and the hull [34] to the miniaturization of this technology to be used in small AUVs [35].

The second type of fixed propulsive technology analyzed here is the jet-based propulsion. Typically, these actuators draw water from an inlet into a pump in charge of adding energy to the fluid. The water then passes through a nozzle, which in turn, transforms the pressure of the liquid in kinetic energy [27, 36], creating a net thrust force used to control the robot.

Jet propulsion offers several advantages over typical thrusters [18, 27], some of these are:

- Mechanical design (absence of rotating parts and transmission mechanisms)
- Cost
- Robustness
- Safety (less likely to cause harm at low speed than thrusters)

However, with regard to traditional thrusters, water-jet actuators are less efficient.

Even if this technology is not as widespread as thruster technology, a non negligible amount of research is devoted to model and improve the characteristics of this type of actuators [36–39]. This will probably lead into efficiency improvements and eventually a bigger adoption of this technology.

9.2.2.2 Vectorial Propulsion

Vectorial propulsion is a technology capable of creating an orientable propulsive force. A vectorial force actuators is then a device in which we have to control no only the generated thrust but also the orientation of said force.

Vectorial propulsion offers interesting possibilities. The first one is the generation of thrust over directions that can change during the AUV operation. Using the reorientation we have access to more degrees of freedom per thruster than a classical fixed propulsive arrangement. If used correctly, vectorial propulsion can provide an underactuated AUV similar capabilities than a fully actuated one. This means giving the possibility to an underactuated vehicle to follow trajectories that otherwise would have been unreachable [40].

Overall AUV weight reduction is a second advantage of this technology. Indeed, given that one of these actuators can be in charge or controlling different degrees of freedom, a reduced number of devices are needed to control the movement of the AUV. A reduced number of actuators results in a reduced AUV weight.

Vectorial propulsion is, however, not exempt of drawbacks and limitations. From a fundamental point of view, vectorial propulsion generate new challenges in control and trajectory planning of AUVs, such as the increase of nonlinearities in the propulsive model [41] and coupling between its actuated degrees of freedom [42]. The versatility of these actuators, when not properly assessed, can be detrimental to the controllability of underwater robots.

From a practical point of view, the biggest limitation is the way of creating the reorientation. Several approaches exist, using different technologies such as water-jets [43, 44], creative mechanism [45], servomotors assemblages [46] and magnetic couplings [47]. However, most of the current solutions are not yet able to match the simplicity of the fixed propulsion approach.

As for water-jet propulsion, it does not mean that development of these actuators has stalled. Improvements and studies about their design and performance are conducted [48], making this a promising technology for the years to come.

9.2.2.3 Control Surfaces

Historically, the first applications of AUV involved performing survey-style tasks. The classic AUV configuration for these missions was a torpedo-shaped hull, a propeller in the rear of the structure and control surfaces [49–51].

In this type of propulsion, the propeller and control surfaces act together in order to propel and steer the robot. The propeller, which is a device consisting of several rotating blades, accelerates the water creating a propulsive force. The control surfaces are in charge of controlling the thrust generated by the propeller, they are airfoil shaped devices capable of generating a lift when positioned at an angle with regard to the flow [52]. The lift depend on the shape of the control surface, its angle and evidently the velocity of the water. This last element is the link between the control surface and the propeller: in order to generate lift, the robot needs to have a non zero velocity.

The propeller and control surface solution was used on ships before their application to AUVs [27], which means that the technology was mature enough when it was implemented on underwater robots. This made the application of this technology in survey-type robots easier, given their efficiency and proven efficacy in long range and high speed missions.

Given the success and popularity of survey style AUVs, the propeller and control configuration was studied extensively [53–55]. Aspects as blade shape optimization [56], interaction between propeller and hull [57] as well as the control surface optimization [58] were thoroughly analyzed.

When AUVs missions started to include middle and short range operations, the limitations of this technology arose. Indeed, given that the steering of the vehicle is only possible when the velocity of the water (with regard with the robot) was high enough, operations with low velocities became inefficient with this technology.

Nowadays, this technology is used almost exclusively in long range AUVs [59] and in underwater vehicles with both long and short range applications [52].

9.2.2.4 Biomimetics

The biomimetic approach proposes to base the AUV development in solutions found in the nature. This means adopting shapes, propulsion systems and locomotion control methods used by animals.

The basic idea behind this design approach is that after eons of evolution, nature has found efficient and reliable locomotion solutions. Biomimetic relies on the fact that the work of creating, testing and evolving, in this case, propulsion systems, has been done already in a natural way [60].

In nature we can find fishes and other aquatic beings which are highly maneuverable, power-efficient and capable of performing demanding tasks such as station-keeping in presence of perturbations [61]. Clearly, the understanding of these systems can be used to inspire artificial human-made underwater systems.

Biomimetic design, as a method for underwater vehicles, covers robot shape, propulsion system and control (locomotion behavior control). In virtue of this, biomimetics can be included in any other section of this chapter. However, given that the final goal for biomimetic design in AUVs is to emulate animal locomotion methods, it is included in this section.

Several examples of biomimetic AUVs exist in the literature, there are robots copying the locomotion methods of squids [62], eels [63], turtles [64], dolphins [65] and fish [66]. Even if the animals inspiring the designs are diverse, most of the biomimetic robots nowadays use (in some way or another) fin propulsion as a base [67, 68].

This technology has generated great interest in the scientific community, and as such, extensive analysis about fins hydrodynamics [69] and real prototypes have been carried out [70, 71]. Another proof of the popularity of biomimetics is the diverse methods to achieve animal-like movement. Indeed, in the literature we can find complex and creative mechanisms [63, 65] as well as solutions using smart materials [72]. These materials have special characteristics that, when applied correctly, allow biomimetic AUVs to perform complex movements. Evidently, biomimetic locomotion also supposes the use of adapted control techniques in order to master the movement of the robot, we can find several examples of these control methods in the literature [73–75].

9.2.3 Conclusion

To finish the state of the art, different AUV design approaches will be discussed. The AUV being a complex mechanism with diverse subsystems from very different domains (hydrodynamics, control, localization, etc.), needs to be designed attending several constraints and specifications. Some of these are counter-productive between them. For instance, the need of a large interior volume to store the payload can result in an AUV with a mediocre hydrodynamic behavior due to drag forces.

Naturally, given the need to reconcile the diverse aspects of AUV design, different techniques and protocols have been used to obtain robots capable of attending all needed characteristics.

The first type of design method used to build AUVs has been sheer experience and intuition. Engineers with previous experience in design of robots and aquatic devices took (and take still nowadays) the daunting task of adjusting the several needed AUV characteristics. For some AUVs, methods, ideas and techniques derived from ship design were borrowed.

The great amount of heterogeneous design parameters and the difficulty of creating efficient AUVs based solely on experience, made optimization based design appealing.

Early optimization methods involved analyzing the risk of failure of the AUVs components and using a computerized evaluation tool to evaluate the need of redundant devices [76]. With advancements in computing power, CFD (Computational Fluid Mechanics) techniques were used to optimize the shape of the AUV, in order to obtain vehicles with low drag and capable of operating at high pressures [77]. Also for reducing drag, other optimization approaches have been taken, such as pattern search [78] and genetic algorithms [79, 80].

In order to take into account the different design parameters intervening in the AUV, multidisciplinary design optimization [81] techniques are also used. These methods use individual analysis for each subsystem and then aggregate them using a system-level coordination procedure [82]. Similarly, multi-objective optimization techniques using genetic algorithms are used with the goal of optimize design parameters in order to obtain better effectiveness, and reduce cost and risk [83, 84].

From this review of underwater robot design, we can see how vast is the spectrum of possibilities in order to optimize the AUVs sub-system. Indeed, given the relatively young age of AUV technology, most of these design are starting to gain maturity, offering exciting possibilities.

The great amount of options, however, implies that we have to carefully choose the set of parameters to use in the optimization of the propulsive capabilities of AUVs.

Each chosen set of parameters could unveil possibilities unlikely to find otherwise. For instance, the study of biomimetic design takes us to the study of the fundamentals of sea creatures locomotion, something that we probably would not do if we study the classical AUV approaches.

In this work, the design of the following sub-systems have been studied:

- Fixed and vectorial thruster technology: The former is an extensively used and established technology. The latter is a promising technology, capable of extending the propulsive capabilities of the AUV.
- Propulsive topology: Given its deep influence in the AUV capabilities to perform complex tasks. Propulsive topology includes the type of actuation (fully or under-actuated), the number of thrusters and their arrangement on the robot.
- Control technique: For the sake of feasibility, a single nonlinear control technique will be used and its parameters are optimized. This is motivated by its adaptation to dynamic tasks, in which linear controllers may have a mediocre performance. Our study consist in adapting this control to different propulsion systems.

Our strategy is to use an evolutionary algorithm to optimize the design of these sub-systems. This choice is based on the good results of this technique and its capabilities of optimizing systems with heterogeneous parameters. For evaluation of robot designs, the choice have been made to rely on dynamic simulation. This can be justified by the complexity to asses the robot performance with regard to the propulsion and control systems. The simulation will show the behavior of the robot on its task without considering its theoretical capabilities, which are hardly assessable.

9.2.4 Principles of Dynamic Reconfiguration

The goal of this kind of problem is given in the following general definition:

Definition 9.1 Global minimum

Given a function f: $M \subseteq IR^n \rightarrow IR$, $M \neq \emptyset$, for $\mathbf{x}^* \in M$ the value $f^* := f(\mathbf{x}^*) > -\infty$ is called a global minimum, iff

$$\forall \mathbf{x} \in M: \quad f(\mathbf{x}^*) \leq f(\mathbf{x}) \tag{9.1}$$

Then, $\mathbf{x}^*$ is a global minimum point, f is called objective function, and the set M is called the feasible region. The problem of determining a global minimum point is called global optimization problem.

In the case of the design of a robot, the global minimum would correspond to the best set of parameters that minimizes (or maximizes) the objective function f. It would correspond to the best robot, based on these parameters.

Generally, the objective function has several *minima* of different values. The optimization can be attracted to follow one of these minima, called *local minima*, even if its value is mediocre compared to other *local minima* or to the *global minima*. We can define mathematically these *local minima* as:

Definition 9.2 Local minimum
For $\hat{\mathbf{x}} \in$ M te value $\hat{f} := f(\hat{\mathbf{x}})$ is called a local minima, iff

$$\exists \varepsilon \in IR^{+*} : \forall \mathbf{x} \in M : ||\mathbf{x} - \hat{\mathbf{x}}|| < \varepsilon \Longrightarrow \hat{f} \leq f(\mathbf{x}) \tag{9.2}$$

A local minima would correspond to a version of the robot which is locally better than its neighboring robots, but not the best of all search space M.

9.2.4.1 Different Types of Methods

There are several classifications of the different global optimization methods. Here, the classification given by Bäck [85] will be used. Indeed, we will separate these methods in two large groups:

- **Volume oriented**: These methods are based on an exhaustive search throughout the entire feasible configurations. This, of course, requires a finite search space. Examples of methods using this paradigm are Monte-Carlo strategies and cluster algorithms
- **Path oriented**: These methods follow a path in the feasible space, starting from a point that can be placed whether randomly or using previous knowledge. Path oriented methods can be divided into two sub-groups:
  - **Prediction methods**: these use an explicit model of the objective function to predict the best path to follow. Tunneling methods and gradient descent are examples of these type of approach.
  - **Exploration methods**: these try different paths and discard the ones that are not useful or successful. Examples of this approach are pattern search, rotating coordinates method and, to a great extent (they have traits corresponding to volume oriented and prediction methods), evolutionary algorithms.

For this work, the evolutionary approach will be used to solve the global optimization problem. This choice is based on the many successful applications of evolutionary algorithms to system design [86–89].

9.2.4.2 Evolutionary Algorithms

Evolutionary algorithms are computational techniques based on the model of natural evolution formulated by Darwin [90]. They model the evolution process of a population of individuals using simplified processes of genetic mechanisms as described by Mendel.

Evolutionary algorithms manipulate the genotype of a population of individuals using genetic operators, this population undergoes a fitness-based selection process, in which the fitness of each individual depends on its adaptation to its "environment". With the proper definition of "genotype", "fitness" and "environment" we can apply these optimization algorithms to a myriad of problems, including robot design optimization [89].

From a mathematical point of view, an evolutionary algorithm can be defined as follows [85]:

Definition 9.3 An Evolutionary algorithm (EA) is defined as an 8-tuple

$$EA = (I, \Phi, \Omega, \Psi, s, \iota, \mu, \lambda) \tag{9.3}$$

where $I = A_x \times A_s$ is the space of individuals, and A_x, A_s denote arbitrary sets; $\Phi : I \rightarrow IR$ denotes a fitness function assigning real values to individuals.

$$\Omega = \{\omega_{\Theta_1}, ..., \omega_{\Theta_z} | \omega_{\Theta_i} : I^\lambda \rightarrow I^\lambda\} \cup \{\omega_{\Theta_0} : I^\lambda\} \tag{9.4}$$

is a set of probabilistic genetic operators ω_{Θ_i}, each of which is controlled by specific parameters summarized in the sets $\Theta_i \subset IR$.

$$s_{\Theta_s} : (I^\lambda \cup I^{\mu+\lambda}) \rightarrow I^\mu \tag{9.5}$$

denotes the selection operator, which may change the number of individuals from λ or $\lambda + \mu$ to μ, where $\mu, \lambda \in IN$ and $\mu = \lambda$ is permitted. An additional set Θ_s of parameters may be used by the selection operator. μ is the number of parent individuals, while λ denotes the number of offspring individuals. Finally, $\iota : I^\mu \rightarrow \{true, false\}$ is a termination criterion for the EA, and the generation transition function $\Psi : I^\mu \rightarrow I^\mu$ describes te complete process of transforming a population P into a subsequent one by applying genetic operators and selection:

$$\Psi = s \circ \omega_{\Theta_{i_1}} \circ ... \circ \omega_{\Theta_{i_j}} \circ \omega_{\Theta_0} \tag{9.6}$$

$$\Psi(P) = s_{\Theta_s}(Q \cup \omega_{\Theta_{i_1}}(...(\omega_{\Theta_{i_j}}(\omega_{\Theta_0}(P)))...)) \tag{9.7}$$

Here $\{i_1, ..., i_j\} \subseteq \{1, ..., z\}$, and $Q \in \{\emptyset, P\}$

Several algorithms use the paradigms of evolutionary optimization, we name here a few:

- **Genetic algorithms**: The most used technique within the evolutionary algorithms. Typically, the individuals are coded using binary encoding and they use crossover and mutation as operators [91].
- **Evolution strategies**: It was born to solve hydrodynamic problems such as shape optimization and drag reduction [92, 93]. In this technique the individuals are coded as vectors of real numbers. It uses primarily mutation, which can be *self-adapting*, depending on the objective function. Genetic operators manipulate random variables and are based on probabilistic formalism.

- **Evolutionary programing**: This technique evolves a population of computer programs to solve an algorithmic problem. The solutions to the optimization problems given by this technique are, then, computer programs [94].

Taking into account their ease of use, the simplicity of their implementation and their popularity (which gives many examples of applicability), the genetic algorithm method will be used from now on to solve the optimization problems in this work. Indeed, task-based propulsive solutions will be created writing the design of an AUV as genetic algorithm.

9.3 Genetic Algorithms

9.3.1 Principle

Being a type of evolutionary algorithm, genetic algorithms (GA) are also based on the evolution theory of Darwing. Genetic algorithms use individuals, genes, populations, a fitness function and operators such as selection, crossover and mutation [91].

On a typical genetic algorithm application, the method follows these steps:

1. A population of individuals is created. This population can be created randomly or using previous information.
2. The individuals of the population are evaluated and given a fitness according to their performance based on the objective function.
3. Using the fitness of each individual as information. Genetic operators are applied to create new individuals:
   a. Selection: The best individuals (best fitnesses) are selected from the population. This selection can be deterministic or stochastic.
   b. Crossover: Two individuals (parents) mix their genes in order to create two new individuals (children).
   c. Mutation: A change of value of the genes (alleles) of the individuals is applied. The change (from 0 to 1 or conversely) is decided by stochastic methods (each gene has a tiny probability of being mutated)
   d. Population replacement: The new generation created by the previous operators replaces completely the old generation.
4. Go to step 2 if solutions are not yet satisfactory. Otherwise, finish the algorithm

9.3.2 Components

9.3.2.1 Population Initialization

The first step using a genetic algorithms is the creation of an original population. This population will contain the first individuals, that later on, will be evaluated and affected by genetic operators.

The μ individuals of a population are typically coded using k binary genes. The creation of these binary genes is made, most of the time, randomly. Indeed, a binary random variable is sampled $k \cdot \mu$ times in order to create every single gene of all the individuals of the population. The process is equivalent to a coin toss. The result is the start population $P(0) = \{\mathbf{a}_1(0), ..., \mathbf{a}_\mu(0)\}$.

Other techniques of population initialization exist. Some of them place individuals in zones of the population were the optimal solution is likely to be found. This type of strategy requires previous knowledge about the optimization problem and may lead to local optimization if mishandled.

9.3.2.2 Evaluation

Once created, the individuals of the population are evaluated. In order to do this, each individual must be decoded from binary to decimal base first.

The decoding is made taking the integer value from 0 to $2^k - 1$ of each individual and then mapping it to a real interval $[u_i, v_i]$. An example of this conversion is given here:

$$\Upsilon^i : IB^{k_x} \rightarrow [u_i, v_i] \Upsilon^i(a_{i1}, ..., a_{ik_x}) = u_i + \frac{v_i - u_i}{2^{k_x} - 1} \left(\sum_{j=0}^{k_x - 1} a_{1(k_x - j)} 2^j \right) \tag{9.8}$$

The decoded individuals are then, one by one, evaluated using an objective function $f(\Upsilon(\mathbf{x}))$ based on the characteristics to enhance/decrease. This objective function is scaled to the domain of the positive real numbers, since the selection operator needs positive fitness to work. This is done with a scaling function δ:

$$\delta : IR \times \Theta_\delta \rightarrow IR^+ \tag{9.9}$$

There are several scaling methods [91] (linear, logarithmic, exponential, etc.). The choice of the right method depends on the application.

The scaling function gives to each individual in the population a *fitness* F, which describes their adaptation to the environment and it is useful to apply the operators effectively.

9.3.2.3 Selection

Selection is the genetic mechanism from which the best individuals of the population are saved from extinction.

The selection is based on the survival of the fittest rule of Darwin. In this type of selection, called *proportional selection*, each individual has a *selection probability* (also called *survival probability*) proportional to its relative fitness. It means that the better fitted an individuals is, with regards to the rest of the population, the bigger the chances of its survival are.

Mathematically we can determine this probability for the individual $\mathbf{a}_i(t) \in P(t) = \{\mathbf{a}_1(t), ..., \mathbf{a}_\mu(t)\}$ as:

$$p_s(\mathbf{a}_i(t)) = \frac{F(\mathbf{a}_i)}{\sum_{j=1}^{\mu} F(\mathbf{a}_j(t))} \tag{9.10}$$

Evidently this calculation only works if the values of the individual fitness are positive.

Using the above defined probability and the number μ of individuals in the population, we can calculate the expected number of occurrences of individuals $\mathbf{a}_i \in P(t)$ in the population after the selection:

$$\eta(\mathbf{a}_i(t)) = \mu \; p_s(\mathbf{a}_i(t)) \tag{9.11}$$

The proportional selection mechanism is materialized by a method called the "roulette wheel". In this digital roulette we assign to each individual a slice of a size proportional to its relative fitness. Then a random number is generated and the individual whose segment spans the random number is selected (Fig. 9.1). This process is repeated for each individual of the population.

9.3.2.4 Crossover

The crossover, considered as the most important operator [85], is the operator by which new individuals on a population are created from parent individuals of the previous population. Indeed, new individuals can be yield from useful elements of different parents. These new individuals have the combination of the characteristics of their ancestors.

The crossover operator selects two parents from the population and recombines their genes (binary data) to produce two new individuals. A pair of individuals has a probability p_c of being crossed. The value of this probability depends on the application. In the literature we can find authors proposing p_c values ranging from 0.5 to 0.95.

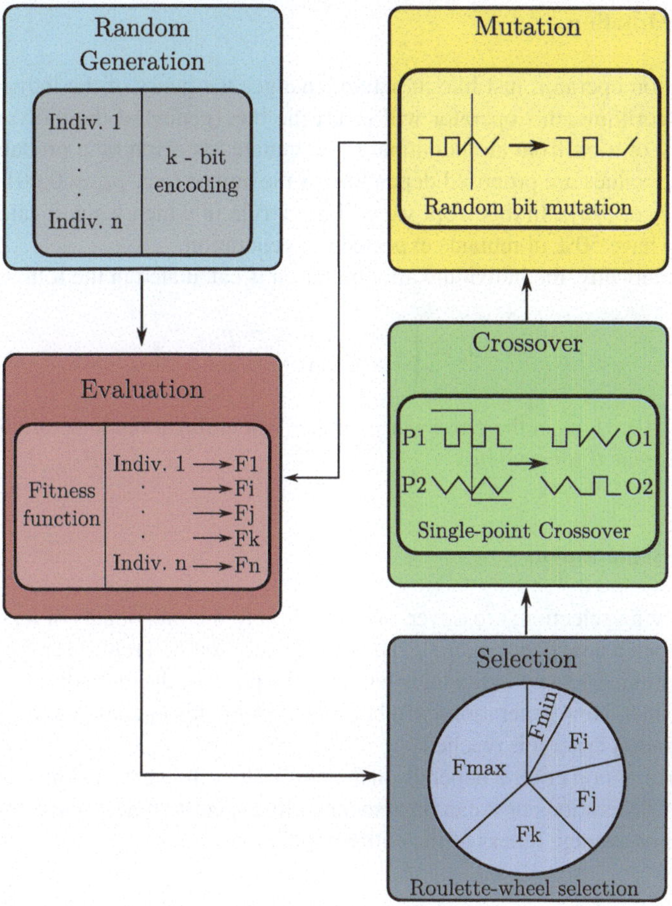

Fig. 9.1 Genetic Algorithm

Another topic of discussion is the crossover technique, that is, the way in which the elements from the parents are combined to form their children. Several techniques exist [85, 95]. In Fig. 9.1 we can see the most basic (and popularly used) technique: the single point crossover.

The single point crossover method, proposed by [96], uses a crossover position $\chi \in \{1, ...k-1\}$ which is selected randomly. The two resulting children are then created exchanging the bits to the right of χ between the parents. From two individuals **s** and **v** the resulting children are:

$$\mathbf{s}' = (s_1, ..., s_{\chi-1}, s_\chi, v_{\chi+1}, ..., v_k) \quad (9.12)$$

$$\mathbf{v}' = (v_1, ..., v_{\chi-1}, v_\chi, s_{\chi+1}, ..., s_k) \quad (9.13)$$

9.3.2.5 Mutation

The mutation operator, just like in nature, changes the genes of the individuals. In genetic algorithms, this operator will invert the bits (genes) of the individual. The probability of a bit from an individual to be mutated is given by a probability p_m. Several p_m values are proposed depending of the author (e.g. $p_m = 0.001$ for [97], $p_m = 0.01$ for [98]). In this work we will use a rule in which we will calculate p_m in order to have 50% of mutants expected per generation.

Mathematically, the individual after mutation is calculated in the following way:

$$a_i' = \begin{cases} a_i & , \text{if } \chi_i > p_m \\ 1 - a_i & , \text{if } \chi_i \leq p_m \end{cases} \tag{9.14}$$

where $\chi \in [0, 1]$, as in the crossover operator, is a uniform random variable which value is recreated for each bit.

9.3.2.6 Replacement

After applying selection, crossover, and mutation to the individuals of a population P_0, we obtain a new population P_1. This new generation will replace completely the old generation. The new individuals will take the place of the old individuals, which will be extinct. New generations will be created using this operation iteratively until the termination criteria is reached.

The termination criteria depends on the application. It can be a simple maximum number of generations or it can be associated to a specific fitness value for the best individual or average fitness of the entire population.

9.3.3 Application Example

9.3.3.1 Rastringin Function

Before applying the genetic algorithm to our particular problem, we need to verify its efficacy. This verification is necessary from a theoretical and practical point of view. Theoretically, it will allow us to confirm the pertinence of these type of algorithm and will improve our grasp over the method. From a practical point of view, it will allow us to understand the logic behind our code.

These type of verification are typically done using a known multimodal multidimensional function, known as the test function. These test functions have the property of having a great number of *local minima* and a known *global maxima*, which makes them a challenging problem, perfect to test the genetic algorithm.

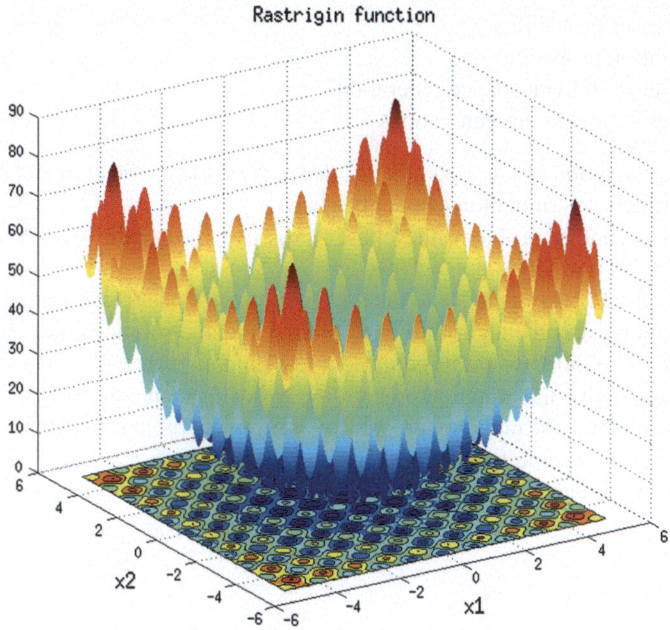

Fig. 9.2 Rasringin function with $n = 2$ (2D)

Among the several functions found in the literature [99], in this work we will use the Rastringin function. This function is multidimensional and highly multimodal, with a large amount of *local minima* surrounding the *global maxima*. Mathematically we can define this function as follows:

$$f(x) = 10n + \sum_{i=1}^{n}[x_i^2 - 10\cos(2\pi x_i)] \tag{9.15}$$

The limits of this function are restricted to the hypercube $-5.12 \geq x_i \leq 5.12$, $i = 1, ..., n$. The *global minima* of this function is $f(x) = 0$, which is the result of $x_i = 0$, $i = 1, ..., n$.

In Fig. 9.2 we can see a plot for the Rastringin function with $n = 2$.

For our test,we will use a Rastringin function with $n = 20$. The configuration of the genetic algorithm is given next.

9.3.3.2 Parameters

In order to use the genetic algorithm, first we have to configure it correctly. The configuration of the algorithm implies the definition of the following parameters:

- pc: crossover probability
- pm: mutation probability
- μ: number of individuals per generation
- g: number of maximum generations

The determination of these parameters depends on the application and a general method for their determination can not be stated.

For this particular problem, the chosen crossover probability will be the one proposed by [91], which is $pm = 50\%$.

The evolution is very sensible to the mutation effect, since a large mutation probability would prevent the algorithm from finding suitable solutions. For instance, a probability $pm = 50\%$ would be equivalent to a random walk.

In order to define the mutation probability, we will use the method used in [87]. Using these method, the mutation probability is inversely proportional to the number of bits of a chromosome (gene arrangement). If we want at least 50% of the total population in average to mutate, then the mutation probability is given by:

$$pm = \frac{1}{2\,n\,k} \tag{9.16}$$

with n the number of variables to optimize and k the number of bits coding these variables.

The number of individuals per generation (μ) in a population must be sufficiently large so the selection operator can pick from a great variety of individuals. Just like in nature, selection is more efficient if it has several type of individuals to chose from. This can lead to make the algorithm be able to "explore" more solutions and "discover" areas of the feasible landscape M that otherwise wold have been hidden.

A way of choosing the value μ can be based on the fact that it should allow the possibility of having the full range of values attainable by the fitness function F:

$$\Omega(F(M)) = k_{min} = 1 + \frac{1}{F_e} \tag{9.17}$$

with F_e the smallest possible variation on the fitness function, i.e. the variation on the fitness function given by a change of the last significant bit in a variable.

To determine the number of generations g we need to take into account that it has to be sufficiently large so the evolution has enough time to find suitable solutions and to improve the fitness of the population. We can chose the number of generations in order to give time to the mutation of modifying the complete chromosome of an individual. Given that the expected number of mutations per generation over an individual is 0.5, the minimum number of necessary generations is:

$$g_{min} = 2\,n\,k \tag{9.18}$$

Table 9.1 Evolutionary parameters for Rastringin optimization test

Parameter	Symbol	Numerical value
Crossover probability	pc	0.5
Mutation probability	pm	1.786×10^{-3}
Population size	μ	1290
Number of generations	g	560
Number of parameters	n	20
Enconding bits per parameter	k	14

Lastly, we can use the previous parameters to check the number of total evaluations on our genetic algorithm with regard to the cardinality of the optimization problem. Indeed, we can evaluate the result of the following equation:

$$\frac{\mu\, g}{\Omega(M)} \tag{9.19}$$

with M the set of solution candidates and $\Omega(M)$ the cardinal of the set M. The smaller the result of this equation, the bigger the difference between the evaluations made by the GA and the number of possible solutions of the optimization problem.

For our test, we will use $n = 20$ (number of variables) and $k = 14$ (bit for the encoding of each variable). This allows us to create the Table 9.1.

9.3.3.3 Results

Using the genetic algorithm with the given configuration we find the following results. In Fig. 9.3 we can see the evolution of both the best and average fitness for the population. It is clear how the average fitness curve increases along the best fitness curve. This shows how the genetic algorithm takes the information from the parents in order to create better and more fitted children as the generations go by. The gap between the two curves, created by the difference in grow rate, is an evidence of the variety of solutions the algorithm maintains during the evolution. Even if some of these solutions are not as good as the best one, they may contain useful information that, thanks to the crossover operator, can lead to improve the best fitness later on. This gap, mostly maintained by the mutation taking place on the individuals of the population, is very healthy for the evolution. Indeed, in its absence, the population converges to μ copies of a single individual. At this point, fitness improvement would become rather difficult and unlikely, since it is only driven by mutation.

Figure 9.4 shows the evolution of the parameters of the best individual. We can see how in the beginning the best solution has parameter values far from zero and how, as the evolution progresses, the values of the variables decrease as new areas of M are explored. The process continues, passing through parameters corresponding to better *local minima*, until the *global minima* configuration is found. At this point

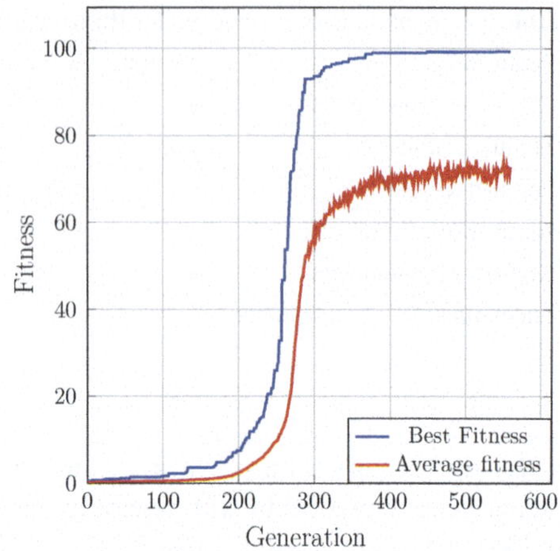

Fig. 9.3 Best and average fitness evolution for Rastringin optimization problem

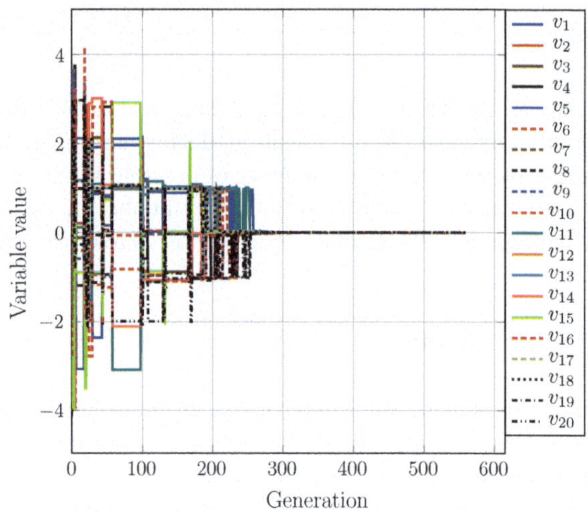

Fig. 9.4 Evolution of the n variables of the Rastingin optimization problem

no further improvement is possible, but the genetic algorithm continues its research until the stop condition is attained.

We can also evaluate the efficiency of the GA by calculating:

$$\frac{\mu\, g}{\Omega(M)} = 3.7186 \times 10^{-19} \tag{9.20}$$

which means that the number of evaluations made by the algorithm was negligible when compared with the possibilities of the research space M.

9.4 Adaptation to AUV Propulsion Design

An underwater robot, regardless of its category, is a complex system affected by various and heterogeneous parameters. Indeed, a small change in these parameters can greatly influence the overall performance of the robot. They can even be the difference between a working and a defective robot.

Considering this, the use of a genetic algorithms to find appropriate parameters to create a suitable robot for a given task seems reasonable.

In this section the adaptation of genetic algorithms to our specific underwater robot optimization problem will be discussed. Descriptions about the configuration of the evolution (probabilities, number of generations, number of individuals, etc.) and of the design parameters (type, encoding, initialization, etc.) will be given.

9.4.1 Design Parameters

Among the many design elements of an AUV, this work will focus on the ones regarding the propulsion and the control. Namely:

- Propulsion parameters
  - Number of thrusters: n
  - Thruster position: P_{ix}, P_{iy}, P_{iz}
  - Thruster Orientation: θ_i, ψ_i
- Controller FBLN parameters
  - Control gains: Λ, K_p
  - Tracking point: P_{ex}, P_{ey}, P_{ez}

Where i represents the ith thruster in the topology.

9.4.1.1 Propulsion Parameters

Propulsion parameters have a deep influence on the operation of underwater robots. The three propulsion parameters listed above will determine the controllability of the underwater vehicle. each new propeller will contribute to the robot controllability with a propulsive wrench. The wrench contribution depends on the position and orientation of the thruster, since these parameters will dictate what DOF will be actuated by the thruster. For each additional thruster we use on the robot, we will either expand its propulsive capabilities (controllability) or simply add a redundancy on its DOF. For a robot with few thrusters, each new thruster can easily improve the robot controllability. However, for a robot with many thrusters each new thruster will have a decreasing contribution on the controllability.

To follow a given trajectory, the necessary DOF can be determined using kinematic considerations. In our genetic algorithm, this will be used to know the minimum propulsive capabilities needed by a robot. This pre-analysis provides an answer to *What* the robot needs in terms of DOF, while the genetic algorithm is in charge of finding *How* to provide them.

The propulsion parameters also have an influence over the energetic performance of the robot. The larger the number of thrusters, the heavier, cumbersome and costly the robot will be. In our genetic algorithm, each additional thruster penalizes the objective function, in order to account for this effect.

Likewise, if the thrusters of the robot are badly placed and oriented, the creation of the propulsive wrench will not be optimal, or worse inefficient, even with an appropriate number of thrusters. A defective propulsion topology needs to generate a greater propulsive wrench in order to follow a defined trajectory. Since our simulation saturates the force on the thrusters (by using the electrodynamical thruster model with bounded voltage), a poorly configured robot could not be as effective to follow said trajectory than a properly configured one. This will translate on bigger tracking errors (and energy consumption) throughout the evaluation.

9.4.1.2 Controller Parameters

Controller parameters also play an important role on the performance of the robot. These parameters have to be precisely tuned in order to make the robot follow the desired trajectory efficiently.

The first parameter, the tracking point e, intervenes in the kinematic part of the controller. It determines the rank of the **T** matrix, which represents the rigid body relation between the velocity of the tracking point $\boldsymbol{\nu}_e$ and the velocity on the mobile frame origin $\boldsymbol{\nu}$. The **T** matrix has to be inverted in order to calculate the necessary velocity of the robot in order to follow the desired trajectory. This means that the position of the tracking point has to guarantee the invertibility of the **T** matrix after the space reduction. Additionally, since ultimately it is the tracking point that follows the desired trajectory, its position will determine as well the convergence seed. Indeed, a tracking point far from the origin of the mobile frame will need less thrust power to correct its trajectory. This is due to the fact that small changes on the robot angular velocity result in big changes on the tracking point linear velocities. The position of this point relatively to the robot may be dictated as well by the task itself.

Lastly, the gain parameters of both the kinematic and dynamic control layers will also influence the robot ability to follow the desired trajectory. A fine tuning of these values is necessary to obtain a rapid and efficient convergence to the reference trajectory. On one hand, both gains need to be sufficiently large to guarantee a rapid convergence to the followed path. However, on the other hand, the kinematic gain also needs to be small enough to create a feasible reference velocity for the dynamic control layer (because of acceleration limitations). Moreover, the dynamic gain need to be small enough to avoid creating chattering on the propellers during the dynamic controller correction.

9.4.2 Representation of Solutions

To properly implement the genetic algorithm, we need to precisely define the design parameters. These parameters are the building blocks used by the genetic algorithm to find adequate solutions to our problem, and they will become the solution at the end of the evolution process.

The parameters being of different nature (we have parameters of position, orientation and others representing controller gains), we need to determine each one on its own domain.

9.4.2.1 Number of Thrusters

The topology of the underwater robot must be optimized, this means that the number of thrusters is variable in the search. To account for this characteristic we will use an activation bit. The activation bit will be in charge of activating or deactivating a thruster, i.e. including or excluding the thruster from the topology.

9.4.2.2 Thruster Position

The thruster position parameter is in reality a position vector, composed of three coordinates expressed in the robot frame: $P_{i\,x}$, $P_{i\,y}$, $P_{i\,z}$ (with i the number of the thruster).

Since the robot has a fixed shape and size, these three coordinates have different limits: $P_{i\,x}$ can take values from $[-L,\ L]$ and $P_{i\,y}$ and $P_{i\,z}$ take values from $[-r_c,\ r_c]$. The volume described by these three coordinates is a cuboid, this corresponds with the real RSM Robot (Fig. 9.5) with its metallic frame. To avoid this heterogeneity we can make these parameters dimensionless:

$$\begin{aligned} \tilde{P}_i = (\tilde{x}_{pi}, \tilde{y}_{pi}, \tilde{z}_{pi}) &\in IR^3_{[-1,1]} \\ \tilde{x}_{pi} &= \frac{2\,P_{i\,x}}{L} \\ \tilde{y}_{pi} &= \frac{2\,P_{i\,y}}{r_c} \\ \tilde{z}_{pi} &= \frac{2\,P_{i\,z}}{r_c} \end{aligned} \tag{9.21}$$

with L and r_c the length and radius respectively.

Each one of these coordinates will be represented by a binary word of 5 bits. This representation will allow to identify 32 points on each axis of the robot, as shown on Fig. 9.6, which allows to have a resolution of 3.23% of the total length. The choice of the discretization is a trade-off between accuracy and calculation time. Indeed,

Fig. 9.5 Structure of RSM robot (IRDL-ENIB, Brest Ifrement bassin)

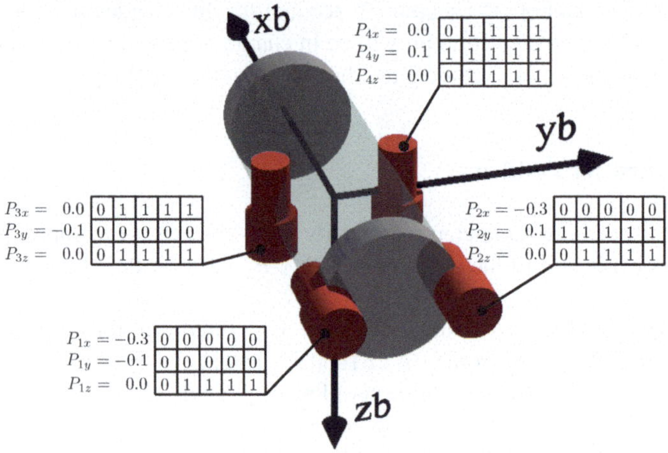

Fig. 9.6 Thrusters position encoding (example of RSM robot)

a small resolution gives a good precision but also increases the search space M, increasing the time necessary to solve the problem.

9.4.2.3 Thruster Orientation

Thruster orientation is a design parameter that contains two sub-parameters: *pitch* angle θ_i and *yaw* angle ψ_i.

These angles can vary within a [-$\pi/2$, $\pi/2$] interval. The variation interval have been shortened since we consider the thruster capable of generating the same forces forward and backwards.

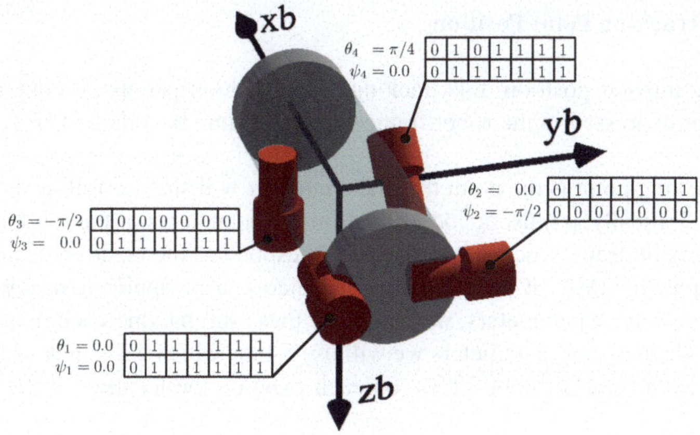

Fig. 9.7 Thruster orientation encoding

As we did with the position of the thruster, we can make these sub-parameters dimensionless:

$$\tilde{O}_i = (\tilde{\theta}_i, \tilde{\psi}_i) \in IR^2_{[-1,1]}$$
$$\tilde{\theta}_i = \frac{\theta_i}{\pi/2} \qquad (9.22)$$
$$\tilde{\psi}_i = \frac{\psi_i}{\pi/2}$$

The encoding of these sub-parameters are made using 7 bits, which allows us to identify 128 possible orientations for each angle, as shown in Fig. 9.7. The encoding, with a resolution of 0.79% of the range, is also a trade-off between accuracy and time of calculation.

9.4.2.4 Control Gains

The control method uses two gains, Λ for kinematic layer and K_p for the dynamic one. For the implementation of the GA we have decided to limit the values of these two sub-parameters to an interval [0, 63.75]. This choice was made based on previous knowledge about the values of λ and K_p, and because it is practical to encode each parameter with 8-bits, which allow us to discretize the variation range in segments of 0.25.

The definition of these two sub-parameters is the following:

$$\tilde{G} = (\Lambda, K_p) \in IR_{[0,63.75]} \qquad (9.23)$$

9.4.2.5 Tracking Point Position

Just as the thruster position, this tracking point position parameter contains three coordinates expressed in the robot frame (with the same boundaries): P_{ex}, P_{ey} and P_{ez}.

The tracking point is the point that the controller will steer to follow the desired trajectory. Typically, it must be located within the boundaries of the robot (cylinder plus its metallic frame), because it should correspond to the position of one of the robot equipments (IMU, SONAR, sensors or camera). In our application, as we did for the thruster position parameters, we will make these sub-parameters dimensionless.

To encode these sub-parameters we will use 5-bits, which will allow us to obtain good precision (resolution of 32.3% on each axis) on the location of the tracking point.

Mathematically, this design parameters is defined as:

$$\begin{aligned}\tilde{P}_e = (\tilde{x}_e, \tilde{y}_e, \tilde{z}_e) &\in IR^3_{[-1,1]}\\ \tilde{x}_e &= \frac{2\,P_{ex}}{L}\\ \tilde{y}_e &= \frac{2\,P_{ey}}{r_c}\\ \tilde{z}_e &= \frac{2\,P_{ez}}{r_c}\end{aligned} \tag{9.24}$$

Table 9.2 sums-up the design parameters and their configuration.

9.4.3 Optimization Problem

As it was introduced earlier, our optimization problem "create a suitable robot propulsion system capable of performing a given mission" is not very well defined. Evi-

Table 9.2 Configuration of design parameters

Name	Symbol	Enc. bits	Range	Resolution
Activation bit	a	1	[0, 1]	–
Thruster pos.(x)	P_{ix}	5	[-0.3 m,0.3 m]	0.019 m
Thruster pos.(y)	P_{iy}	5	[-0.1 m,0.1 m]	0.006 m
Thruster pos.(z)	P_{iz}	5	[-0.1 m,0.1 m]	0.006 m
Thruster or. ($\mathbf{y}_b$)	θ_i	7	$[-\pi/2, \pi/2]$	0.025 rad
Thruster or.($\mathbf{z}_b$)	ψ_i	7	$[-\pi/2, \pi/2]$	0.025 rad
Kin. gain	Λ	8	[0, 63,75]	0.25
Dyn. gain	K_p	8	[0, 63,75]	0.25

dently, a mathematical and concise definition of our problem will allow us to correctly design and implement the genetic algorithm.

The controller calculates the force to generate on the thrusters of a specific robot in order to make it follow a desired trajectory. We can also consider that a high level mission is a succession of properly tunned trajectories chained one after the other. Taking into account these two statements we can conclude that our controller, by making the robot follow the desired trajectory, also makes the robot comply with a given complex mission.

The optimal robot for the optimization problem must be one that follows precisely the desired trajectory given to the controller. This means that the distance from the tracking point to the desired trajectory must reach to zero and stay like that. We can rephrase the last statement saying that the integral of the trajectory error must be as close to zero as possible. Our optimization problem is then, the minimization of an objective function defined as the integral of the kinematic error ϵ_{kin} of the controller. This can be put mathematically as follows:

$$\text{Find } \mathbf{x}^{\star} \in IR|_{[-1,1]}^{5n+5} : (\forall \mathbf{x} \in IR|_{[-1,1]}^{5n+5},\ g(\mathbf{x}^{\star}) \leq g(\mathbf{x}) \tag{9.25}$$

with:

$\mathbf{x}^{\star}$ The most suitable individual formed by $5n + 5$ parameters. Its is searched in a discrete subspace of $IR|_{[-1,1]}^{5n+5}$

$g(x)$ The objective function to minimize, defined as the integral of the kinematic error of the controller over the time of the mission.

As seen in the previous equation, the number of parameters forming an individual can change. Each propeller has 5 parameters: 3 position coordinates (x_{pi}, y_{pi}, z_{pi}) and 2 orientations (θ_{pi}, ψ_{pi}), all of these defined with regard to the robot frame. Given that in the general optimization problem we want to find not only the best position and orientation for the thrusters but also an adequate propulsive topology, the number of thrusters, n, is a variable.

Since the robot has only one main controller, we have 5 parameters to optimize. The parameters of the controller are: 2 gains (for the kinematic and dynamic control layers) and 3 position coordinates (x_e, y_e, z_e), for the position of the tracking point e.

9.4.4 Evaluation

Once the optimization problem and the design parameters have been defined, we need to establish how the genetic algorithm will determine whether an individual is fitted or not. This is done through an evaluation phase, in which the performance of an individual is determined by an objective function. The result of the objective function will be then scaled to obtain a fitness function, that will be used by the selection operator of the genetic algorithm.

The evaluation phase is done typically using a dynamic simulation of the real system using a numerical model. This cost effective method allows a great quantity of evaluations in a relatively short period of time. Additionally, it allows to evaluate solutions that might be harmful for the robot in the real world. This type of evaluation method has, however, a very important draw back: the *reality gap*, an issue by which the solutions found using the GA are not efficiently transfered to the real system [100]. This *reality gap* is often created by inaccuracies in the model, which the GA can use to obtain cheated higher fitness.

In the literature we find several attempts to cross the *reality gap*, here we name a few.

9.4.4.1 Simulation Based

Since one of the main reasons of the existence of the *reality gap* are the inaccuracies of the simulations, a straightforward way of solving this problem would be creating more precise simulations. However, the more accurate the simulation is, the more computation-intensive it gets, which increases the time needed to find proper solutions. Additionally, model accuracy improvement has a limit, specially for systems involving fluid dynamics.

Other simulation-based methods propose to prioritize the evolution of solutions presenting a robust behavior. This can be done by means of a reduced simulation taking into account only the meaningful phenomena and dismissing the complex behavior of the real system [101]. As a consequence of this, the genetic algorithm is not able to exploit the inaccuracies of the system, since they are hidden. Robustness is also obtained by evaluating solutions on different simulation environments and initial conditions [102] or by using adaptive mechanisms, such as neural network controllers [103].

However, despite the drawbacks linked to the inaccuracies, the used models generally describe the studied system in a proper way. Most models contain both accurate and inaccurate parts, the task of the GA user is to interpret whether the outcome of these models is appropriate or not.

In our application of genetic algorithms, we will use simulation as a evaluation mean. Our simulator *EAUVIVE*, far from being perfect (lacking of fluid dynamics and experimental validation of added mass terms), offers an acceptable degree of accuracy considering our goal: finding clues that could aid to the design of underwater robots.

9.4.4.2 Reality Based

Needless to say, the simplest way of avoiding the *reality gap* problem is not having it. The reality-based approaches are based on this premise, as they consist on using the real system to evaluate the solutions. Evidently, a direct application of this approach is very time consuming since each evaluation is done in real time on the robot [103–105].

Other methods based on this approach have been tested. Namely a two step method, in which simulation is used first to determine a set of well-fitted solutions and then the evolution is completed on the real system [106, 107].

Since these approaches are based on the use of simulation to narrow down the research, they are subject to the same considerations of the simulation-based methods.

9.4.4.3 Robot-in-the-loop Based

This approach also uses a co-evolutionary process, in which the algorithm evolves both the simulator and the controllers. Using this method, observed real data has to be correctly described by the simulator, then the best controller used on these simulators is implanted on the real system. This generates a new set of observed real data, which is what the simulator will try to describe on the next iteration [108, 109]. Similar methods [110] based on co-evolution try to reduce the fitness variation between the simulation and the reality.

These methods are based under the assumption that a simulator can be sufficiently improved to allow a perfect transfer between simulation and reality.

Other methods based on the robot-in-the-loop approach use a transferability function [100]. This method, based on fixed non-evolving simulations, makes the evolution aware of the limits of the model. It uses transferability measures that compares the simulated behavior with the real one and calculate its closeness. The transferability measures, as they are not possible to calculate explicitly for the complete research space, is interpolated from a few selected solutions. With this method the fitness and the transferability objective are optimized.

9.5 Dynamic Configuration

The global optimization method allows us to obtain promising results in terms of performance. We can find a suitable robot for a given task, which is our main goal. However, giving the GA the complete liberty of searching the best configuration for each single task could generate impractical solutions (the same robot must perform several task in the same mission). Predicting and optimizing all possible tasks appear to be impossible. The other option is to reconfigure dynamically the robot, which means it will change its propulsion configuration during the mission, while it is under the sea.

Nowadays, propulsion technology is not capable of providing full reconfiguration capabilities. Reconfiguration of the orientation of the thrust seems to be giving its first steps out of infancy, but dynamic (during mission) reconfiguration of the position of the thrusters has not yet received interest.

Taking into account these restrictions, it seems interesting to test the algorithm in a more *realistic* scenario. To do this, we will try to find suitable solutions for the same inspection mission but without moving thrusters (positions are fixed) or

Table 9.3 Evolutionary parameters for dynamic configuration

Parameter	Symbol	Numerical value
Population size	μ	40
Number of generations	g	max 500
Number of parameters	n_p	25
Number of coding bits (thrust. or.)	k_{pos}	5
Number of coding bits (thrust. pos.)	k_{or}	7
Number of coding bits (gains)	k_{gains}	8
Number of coding bits (e. pos.)	k_{e}	5
Size of genotype	s_{gen}	147
Crossover probability	pc	0.5
Mutation probability	pm	5.74×10^{-3}

changing their number (we keep RSM parameters, i.e. 4 fixed thrusters). Only the thrusters orientations and control parameters will be optimized. This solution is realistic because it can be implemented using vectorial thrusters such as the RMCT. We will continue using the *topology factor*, in order to compare the fitness of the dynamic configuration to the one of the global optimization. For these optimizations the evolution parameters are shown in Table 9.3.

9.5.1 Seabed Inspection

This time, the achievement of the optimization will be greatly limited since the GA will only be able to change the orientation of the four thrusters and the control parameters.

However, despite the constraints, the algorithms finds a solution comparable in performance with the global one. The final fitness value is 35.52, as seen in Fig. 9.8. This reduction could be due, evidently, to the fact that less parameters can be changed, but also due to the fact that the number of thrusters is set to four, changing the topology factor to 0.9.

Figure 9.9 shows the simulation and the evolution of the kinematic error. The complexity of the tasks is observed in the evolution of the error. Indeed, we can see how, despite being small throughout the simulation, the error has strong variations during the turning phases of the task.

Figure 9.10 shows the configuration of the robot. We can see that thrusters one, two and four are able to create *yaw*. Which is important to be able to steer the vehicle during the turning phases. We see as well how *surge* is now created by thrusters two and four. Even if they are on the same side, the *yaw* perturbation gets compensated by the orientation of thruster two. Additionally, thruster three gets to act over y_b in order to strengthen the lateral trajectory control.

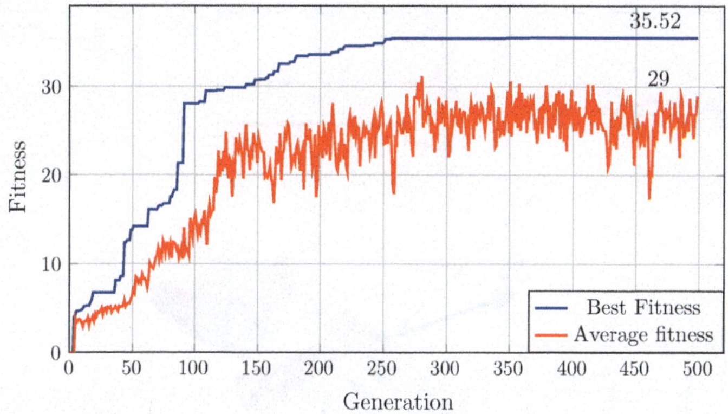

Fig. 9.8 RSM robot fitness for seabed inspection (dynamic conf.)

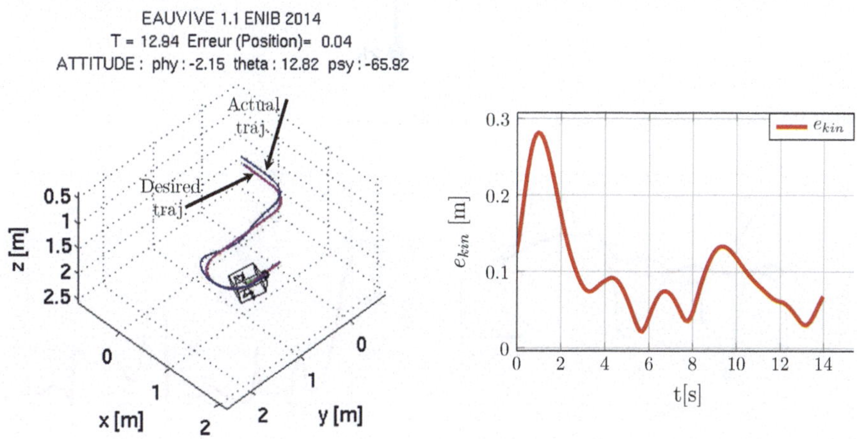

Fig. 9.9 EAUVIVE simulation and kinematic error evolution for seabed scanning phase (Dyn. conf.)

All four thrusters have a component over z_b, allowing them to intervene in the control of *heave* and *roll*. The lack of dedicated thrusters in order to fulfill this duty is due to the small control effort needed to correct these two degrees of freedom for this robot.

The position of *e* is not surprising. The main movement of the robot is to advance, so it is expected to be located in the front of the robot. Allowing to profit from the moment arm from its position to the one of the thrusters.

As with the global controller gains, this time the GA has found a somehow loose kinematic gain ($\Lambda = 1$) and a tighter control over the the dynamic effects on the robot ($K_p = 1.5$).

Figure 9.11 shows that, as in the global optimization for this task, all thrusters have a very active participation in the control of the vehicle. Indeed, we can see how

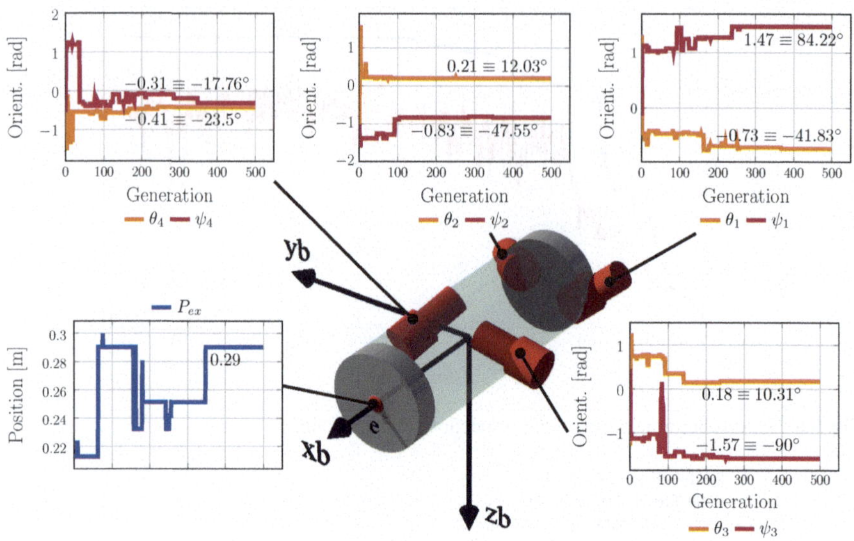

Fig. 9.10 RSM robot for seabed inspection (dynamic conf.)

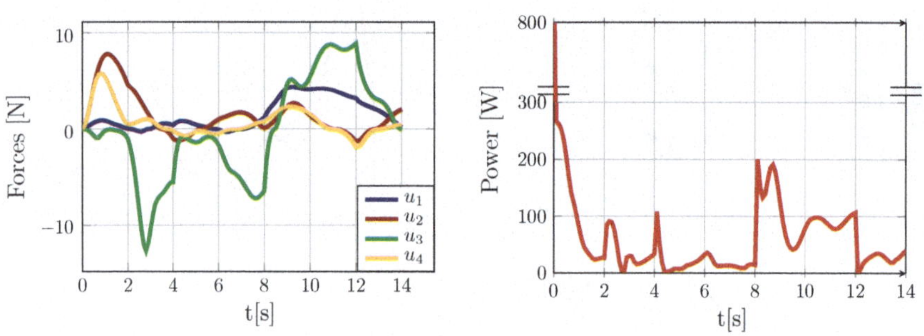

Fig. 9.11 Forces and power consumption during the seabed scanning phase (dyn. conf.)

the thrust forces are constantly variating, specially during the turning phase, which creates a noticeable increment in the the consumed power.

9.5.2 Diving Toward Water Turbine

In this transition phase, the vehicle needs to dive preferably with a *pitch* angle in order to reduce drag forces. This is a type of mission that the RSM robot was not originally designed for. In the global optimization application we have seen that the found configuration involved not controlling the *pitch* angle and just focusing

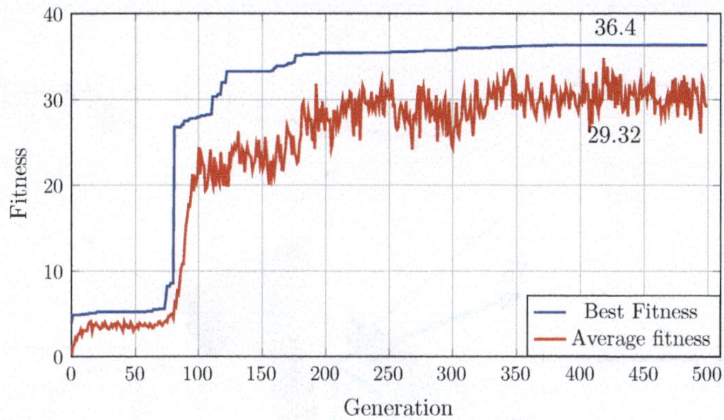

Fig. 9.12 RSM robot fitness for the diving phase (dynamic conf.)

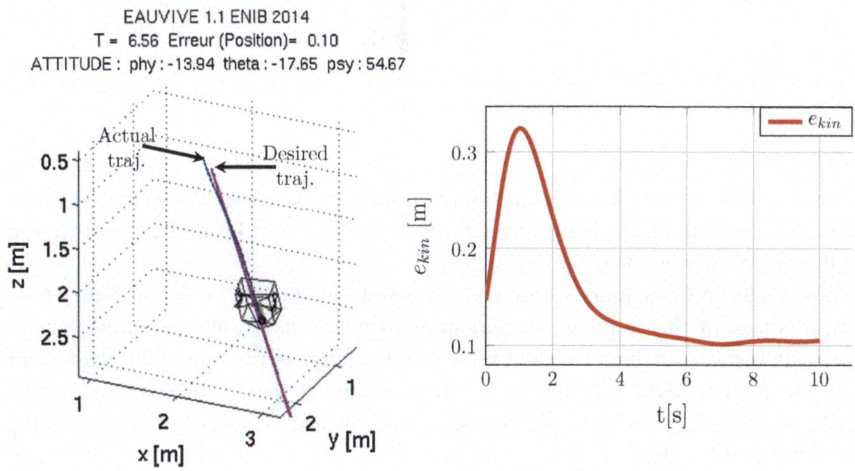

Fig. 9.13 EAUVIVE simulation and kinematic error evolution for diving phase (Dyn. conf.)

in following the position of the trajectory. For this optimization we give again the possibility to the algorithm to find creative ways of performing the mission.

Figure 9.12 shows the fitness of the optimization. Despite the added difficulty of the problem (recreating the same movements than before but with less parameters to modify), we see that the algorithm does not take longer until it begins to consistently improve the individuals of the population.

Simulation and the evolution of the kinematic error are shown in Fig. 9.13. From the kinematic error evolution we can note how the task is straightforwardly executed with minimal variations in the error after the desired trajectory is reached. Additionally, we can also observe that this time the GA has found a solution using *pitch* during the diving.

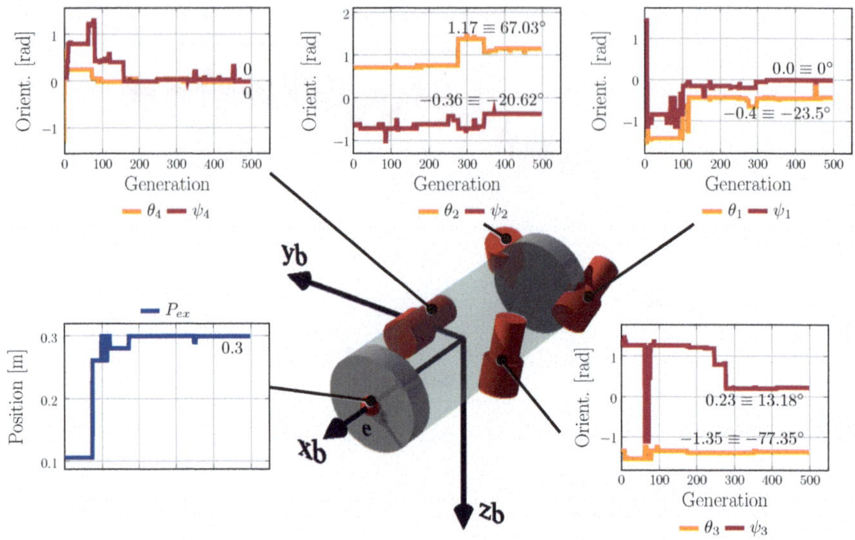

Fig. 9.14 Final robot for diving phase (dynamic conf.)

In Fig. 9.14 we see the solution found by the GA. The four degrees of freedom needed to perform this task are *surge*, *heave*, *pitch* and *yaw*. From the figure we can deduce how the motions are created.

The *surge* and *yaw* motions are created mainly by thrusters one and four. These thrusters take in charge the advancement and lateral correction of the trajectory of *e*. For their part, thrusters two and three are in charge of the remaining degrees of freedom: *heave* and *pitch*, following a similar arrangement as the other two actuators. The inverted configuration of the thrusters gives stability while correctly controlling the degrees of freedom.

The point *e* is positioned in the front of the robot, this gives thrusters one and four a better moment arm to create *yaw*. This position also helps thrusters two and three create *pitch*. Controller gains are $\Lambda = 0.5$ and $K_p = 2.76$, this means that the kinematic controller slowly reacts to the error but makes the dynamic controller rapidly compensate the dynamic effects.

Thrust forces during the simulation and the evolution of the consumed power are shown in Fig. 9.15. We can see how the creation of *pitch* by thruster number two and *heave* by thruster number three are very demanding in terms of force. Indeed, in order to create *pitch*, thruster number two needs to fight against the restoring forces, while thruster three needs to create the diving force. Thrusters one and four are used to give small lateral corrections on the trajectory.

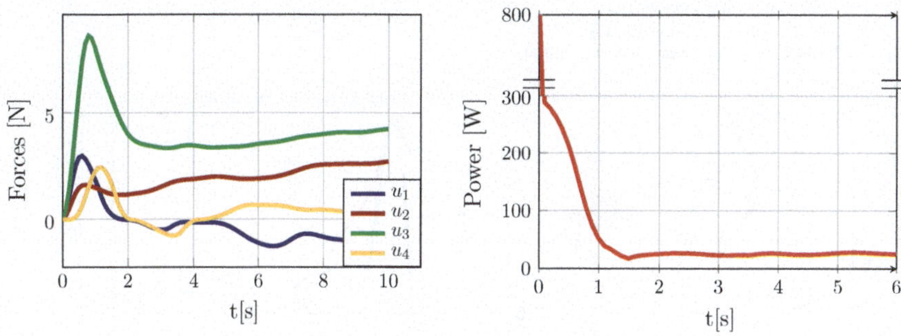

Fig. 9.15 Forces and power consumption during diving phase (dyn. conf.)

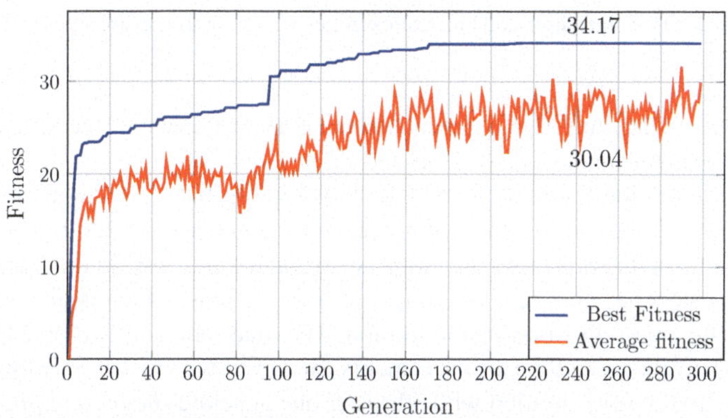

Fig. 9.16 Fitness evolution for tomography phase (dynamic conf.)

9.5.3 *Tomography*

For the last part of the mission, the RSM robot needs to adapt its thrusters to perform a task that, as seen in the global evolution section, involves mostly *sway* and *yaw*.

Figure 9.16 shows the fitness throughout the evolution. We can see how the GA smoothly converged to a solution despite the difficulty of the problem. In fact, if we compare with the result of the global optimization for the same task, the fitness in this case is slightly superior. This is due to the search space reduction, now the genetic algorithm does not loose time evaluating unuseful topologies. In the global optimization, the GA had to first find promising topologies and then optimize their configuration. The fact that the global approach found a four thrusters topology shows how, by reducing the problem to the RSM robot configuration, we have spared the GA of great part of the search effort (less time used for exploration and more exploitation).

Figure 9.17 shows a snapshot of the simulation and the evolution of the kinematic error. As for the global optimization of this task, the robot converges slowly but with

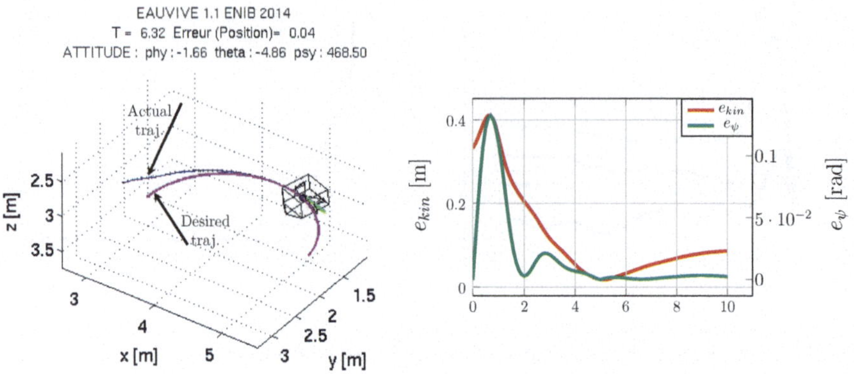

Fig. 9.17 EAUVIVE simulation and kinematic error evolution for tomography phase (Dyn. conf.)

a steady pace. Once the AUV has reached the desired trajectory, the kinematic and *yaw* errors remain small ($e_{kin} < 0.1m$ and $e_{\psi} < 1.4°$).

The configuration found by the GA is shown in Fig. 9.18. As stated in the application of the global optimization, the necessary degrees of freedom to control in order to perform this trajectory are *surge*, *sway*, *heave* and *yaw*. In the figure we see how all thrusters have a force component over the y_b axis, which is understandable since it is the main direction of the motion. Thrusters one and two are, like in the RSM robot, in charge of the *yaw* movements. However, we see that only thruster two contributes to the *surge* motion while thruster one generates *heave* and *pitch*. For its part, thruster three generates mainly a lateral force, controlling *sway*. Lastly, *heave* is generated by a composition of the forces over the z_b from thrusters one and four.

The main movements being *sway* and *yaw*, so the position of *e* was to be expected. The position of the tracking point, close to the R_b origin, favors the control over *yaw*. It gives the thrusters in charge of this degree of freedom a long enough lever arm, which is also short enough to limit the *overshoot* due to propeller dynamics. The GA chose again a small kinematic gain ($\Lambda = 0.25$) making the robot slowly converge to the task, however, it rapidly compensates the dynamic effects on the robot thanks to its dynamic gain value ($K_p = 9.28$).

Figure 9.19 shows thruster forces and power consumption throughout the task. As for the global optimization of this task, the strong dynamic compensation of the controller demands full force from the thrusters, generating in the force peaks of short duration. Once the robot has reached the desired trajectory, we see how thruster number three generates the higher force in order to move the AUV laterally. Thruster number two is the second biggest force contributor, taking in charge the creation of *yaw*. Thrusters one and four have a small contribution, they are mostly in charge of controlling depth.

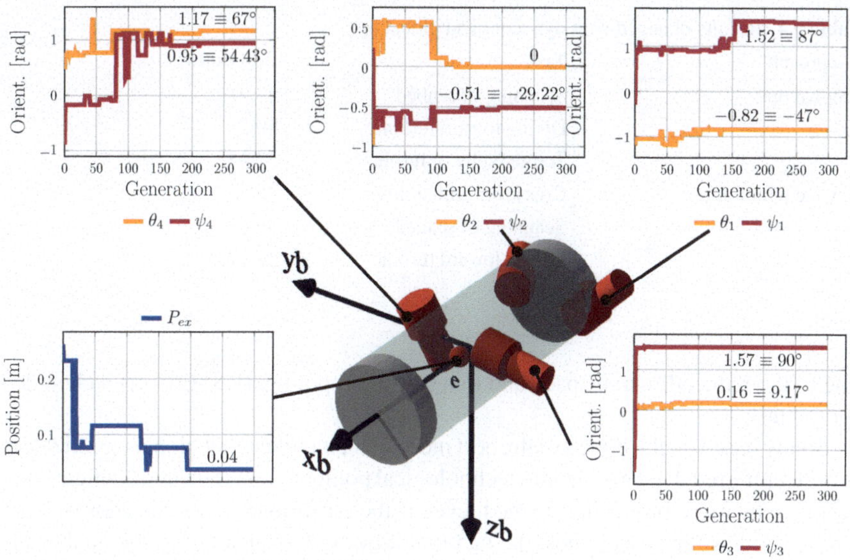

Fig. 9.18 Final robot configuration for the tomography phase (dynamic conf.)

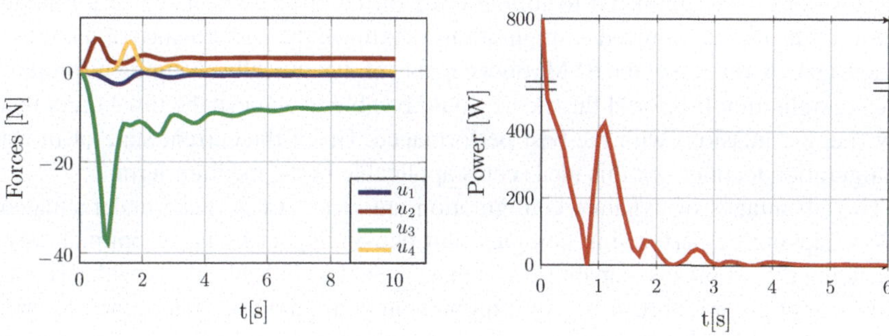

Fig. 9.19 Forces and power consumption during tomography(dyn. conf.)

9.6 Conclusion

Analyzing the results of the dynamic configuration and global optimizations (Table 9.4), we can see how the global approach tends to have better results. For the seabed scanning, the fitness of the best individual found by the global approach was 1.35 times better than the one found by the dynamic configuration optimization. We obtain a similar outcome analyzing the two approaches for the diving task, the global optimization best individual is 1.38 times fitter than the one found by the dynamic configuration. However, for the tomography task, the dynamic configuration finds a better suited robot than its global counterpart (1.05 times fitter), probably

Table 9.4 Results obtained with dyn. conf and global opt

Approach	Task	Best fitness
Global opt.	Scanning of seabed	47.97
	Diving toward turbine	50.2
	Tomography of turbine	32.45
Dyn. configuration	Crossover probability	35.52
	Scanning of seabed	36.4
	Diving toward turbine	34.17

due to the reduced search space and to a starting topology (RSM robot) well suited for the task.

Solely based on the fitness of the best individual, it is clear that global optimization is to be preferred. However from a technological point of view, dynamic configuration seems to be more promising. Indeed, even if the performances of this approach are not as good as the global ones, the fact that they are technically feasible makes the dynamic configuration more useful.

In Sect. 9.5 the RSM robot propulsion system, with predefined thruster number and positions, was optimized to follow three different tasks included in a robotic mission. The three optimized configurations maximize the performance of the robot in each task. A version of the RSM robot capable of dynamically changing its propulsive configuration (vectorial thrusters) would be able to achieve the three tasks in a row (i.e. the mission) with the best performance. Given the current state of thrust configuration technology, this idea seems applicable in the medium-term.

The advantages of dynamic configuration are clear, it can make underactuated AUVs capable of performing complex and demanding tasks in an optimal way, enhancing the propulsive capabilities of these vehicles. Indeed, AUVs with dynamically reconfigurable thrusters, using propulsion configurations determined by this method, would be able to perform tasks with good trajectory tracking convergence and reduced power consumption. This would result in improving AUVs capabilities and therefore expanding their autonomy.

References

1. C. Holden, K.Y. Pettersen, in *IFAC Proceedings*, vol. 40, no. 17, p. 343, 2007. https://doi.org/10.3182/20070919-3-HR-3904.00060
2. P. Herman, Ocean Eng. **36**(6–7), 529 (2009). https://doi.org/10.1016/j.oceaneng.2009.02.003
3. O.E. Fjellstad, T. Fossen, IEEE, J. Ocean. Eng. **19**(4), 512 (1994). https://doi.org/10.1109/48.338387
4. G. Antonelli, *Underwater Robots: Motion and Force Control of Vehicle-Manipulator Systems (Springer Tracts in Advanced Robotics)* (Springer, New York Inc, 2006)

5. L.L. Whitcomb, *Proceedings 2000 ICRA. Millennium Conference. IEEE International Conference on Robotics and Automation. Symposium Proceedings* (Cat. No.00CH37065), vol. 1, p. 1, 2000. https://doi.org/10.1109/ROBOT.2000.844135
6. R. Wernli, in *Proceedings of the 2002 International Symposium on Underwater Technology (Cat. No.02EX556)* (IEEE, 2002), pp. 309–314. https://doi.org/10.1109/UT.2002.1002444
7. B.K. Sahu, S. Member, in *2014 First International Conference on Automation, Control, Energy and Systems* (IEEE, 2014), pp. 1–6. https://doi.org/10.1109/ACES.2014.6808014
8. A.B. Phillips, L. Steenson, E. Rogers, S.R. Turnock, C. Harris, M.E. Furlong, A. Philips, L. Steenson, E. Rogers, S.R. Turnock, C. Harris, M.E. Furlong, Trans. R. Inst. Nav. Archit. Part A Int. J. Marit. Eng. **155**(PART A4), 171 (2013)
9. L.V. Steenson, A.B. Phillips, M.E. Furlong, E. Rogers, S.R. Turnock, S.R. Steenson, L.V. Phillips, A.B. Furlong, M. Rogers, E. Turnock, in *International Undersea Untethered Submersible Technology Conference* (2011)
10. D. Ribas, N.N. Palomeras, P. Ridao, M. Carreras, A. Mallios. Girona 500 AUV: From survey to intervention (2012). https://doi.org/10.1109/TMECH.2011.2174065
11. A. Billings, C. Kaiser, C.M. Young, L.S. Hiebert, E. Cole, J.K. Wagner, C.L. Van Dover, *Deep Sea Res* (Part II Top. Stud, Oceanogr, 2016)
12. K. Alam, T. Ray, S.G. Anavatti, Ocean Eng. **88**, 627 (2014)
13. K. Vestgard, R. Hansen, P.E. Conference, P. Engineers, B. Jalving, O.A. Pedersen, in *ISOPE-2001 Eleventh International Offshore and Polar Engineering Conference*, vol. IV (2001), pp. 679–684\r702
14. R. Marthiniussen, K. Vestgard, R. Klepaker, in *Ocean. '04 MTS/IEEE Techno-Ocean '04 (IEEE Cat. No.04CH37600)*, vol. 2 (IEEE), pp. 846–850. https://doi.org/10.1109/OCEANS.2004.1405571
15. N.A. Cruz, A.C. Matos, in *Ocean. 2008* (IEEE, 2008), pp. 1–6. https://doi.org/10.1109/OCEANS.2008.5152096
16. S. Dupré, G. Buffet, J. Mascle, J.P.P. Foucher, S. Gauger, A. Boetius, C. Marfia, S. Dupré, G. Buffet, J. Mascle, J.P.P. Foucher, S. Gauger, A. Boetius, C. Marfia, T.A.A.Q.R. Team, T.A.A.Q.R. Team, T.B. scientific Party, Mar. Geophys. Res. **29**(4), 275 (2008)
17. V.H. Fernandes, A.A. Neto, D.D. Rodrigues, in *2015 Acoustics in Underwater Geosciences Symposium(RIO Acoustics* (IEEE, 2015), pp. 1–5. https://doi.org/10.1109/RIOAcoustics.2015.7473607
18. A. Alvarez, A. Caffaz, A. Caiti, G. Casalino, L. Gualdesi, A. Turetta, R. Viviani, Ocean Eng. **36**(1), 24 (2009)
19. R. Bachmayer, N.E. Leonard, E. Fiorellit, P. Bhattat, D. Paleyf, J. Graver, E. Fiorelli, P. Bhatta, D. Paley, in *Proc. 2004 International Symposium on Under Water Technology (IEEE Cat. No.04EX869)* (IEEE, 2003), pp. 195–200. https://doi.org/10.1109/UT.2004.1405540
20. N. Stilinovic, D. Nad, N. Miskovic, in *MTS/IEEE Ocean. 2015 - Genova Discov. Sustain. Ocean Energy a New World* (IEEE, Genova, 2015), pp. 1–4. https://doi.org/10.1109/OCEANS-Genova.2015.7271670
21. D. Souto, A. Faina, F. Lopez-Pena, R.J. Duro, in *Proceedings - IEEE International Conference on Robotics and Automation* (IEEE, 2013), pp. 3409–3414. https://doi.org/10.1109/ICRA.2013.6631053
22. P. Ridao, M. Carreras, D. Ribas, P.J. Sanz, G. Oliver, Annu. Rev. Control **40**, 227 (2015)
23. F.Y. Bi, Y.J. Wei, J.Z. Zhang, W. Cao, Control Theory Appl. IET **4**(11), 2369 (2010). https://doi.org/10.1049/iet-cta.2009.0265
24. K. Do, Z. Jiang, J. Pan, H. Nijmeijer, *Proceedings of the 41st IEEE Conference on Decision and Control* 2002, vol. 1, (December), p. 504, 2002
25. J. Refsnes, K. Pettersen, A. Sorensen, *Proceedings of the 41st IEEE Conference on Decision and Control*, pp. 43–50 (2006)
26. M. Chyba, R.N. Smith, *IFAC Proceedings*, vol. 41(1), p. 79 (2008)
27. J. Carlton, *Marine Propellers and Propulsion* (Butterworth-Heinemann, 2012)
28. Y. Tipsuwan, P. Hoonsuwan, in *2015 7th International Conference on Information Technology and Electrical Engineering* (IEEE, 2015), pp. 382–387

29. A. Marouchos, B. Muir, R. Babcock, M. Dunbabin, in *MTS/IEEE Genova: Discovering Sustainable Ocean Energy a New World* (IEEE, 2015), pp. 1–7
30. J. Pyo, H. Cho, H. Joe, T. Ura, S.C. Yu, Ocean Eng. **109**, 517 (2015)
31. A.J. Healey, S.M. Rock, S. Cody, D. Miles, J.P. Brown, IEEE, J. Ocean. Eng. **20**(4), 354 (1995)
32. A. Palmer, G.E. Hearn, P. Stevenson, in *IFAC Proceedings*, vol. 41(1), p. 91 (2008)
33. L. Whitcomb, D. Yoerger, in *Challenges of Our Changing Global Environment. Conference Proceedings Ocean. '95 MTS/IEEE*, vol. 2(5), p. 1019 (1995)
34. A. Saunders, M. Nahon, in *Ocean. '02 MTS/IEEE*, vol. 1 (IEEE), pp. 250–259
35. S. Abu Sharkh, S. Lai, S. Turnock, IEE Proc. - Electr. Power Appl. **151**(5), 513 (2004)
36. Y. Yang, Y. Xie, S. Nie, China Ocean Eng. **28**(3), 409 (2014)
37. U.A. Korde, Ocean Eng. **31**(10), 1205 (2004)
38. A.P. Thomas, M. Milano, M.G. G'Sell, K. Fischer, J. Burdick, in *IEEE International Conference on Robotics and Automation 2005. ICRA 2005* (IEEE, 2005), April, pp. 181–187
39. L. Jian, L. Xiwen, Z. Zuti, L. Xiaohui, Z. Yuquan, Ocean Eng. **123**, 327 (2016)
40. E.P. Vega, O. Chocron, J.V. Ferreira, M. Benbouzid, P.S. Meirelles, in *IECON 2015 - 41st Annual Conference of the IEEE Industrial Electronics Society* (IEEE, 2015), pp. 000,944–000,949
41. S. Jin, J. Kim, J. Kim, T. Seo, in *IEEE International Conference on Advanced Intelligent Mechatronics, AIM 2014*, pp. 1547–1551
42. Y. Le Page, K. Holappa, in *Ocean, MTS/IEEE Conference and Exhibition*, vol. 2000, pp. 2129–2134 (2000)
43. B. Xin, L. Xiaohui, S. Zhaocun, Z. Yuquan, Ocean Eng. **74**, 133 (2013). https://doi.org/10.1016/j.oceaneng.2013.10.003
44. L. Shi, S. Guo, S. Mao, C. Yue, M. Li, K. Asaka, J. Bionic Eng. **10**(4), 446 (2013)
45. E. Cavallo, R.C. Michelini, V.F. Filaretov, J. Intell. Robot. Syst. Theory Appl. **39**(4), 365 (2004)
46. H. Zhang, L. Hao, Y. Wang, Y. Liu, Z. Wu, S. Wang, S. Shao, D. Wei, W. Hou, in *2013 Ocean. - San Diego* (MTS, San Diego, 2013), January 2012, pp. 4–8
47. O. Chocron, U. Prieur, L. Pino, IEEE/ASME Trans. Mechatron. **19**(2), 642 (2014)
48. L.E.J. Ackermann, K.D. Von Ellenrieder, D. Karl, D. Beach, in *Ocean. 2006* (IEEE, 2006), pp. 1–4
49. O. Hegrenaes, O. Hallingstad, B. Jalving, in *2007 Symposium on Underwater Technology. Work. Scientific Use of Submarine Cables and Related Technologies*, vol. 7491 (IEEE, 2007), pp. 558–567
50. J. Bellingham, C. Goudey, T. Consi, J. Bales, D. Atwood, J.J. Leonard, C. Chryssostomidis, in *Proceedings IEEE Symposium on Autonomous Underwater Vehicle Technology*, pp. 148–155 (1994)
51. E. Bovio, D. Cecchi, F. Baralli, Annu. Rev. Control **30**(2), 117 (2006)
52. A.R. Palmer, M. Rights, Analysis of the propulsion and manoeuvring characteristics of survey-style AUVs and the development of a multi-purpose AUV. Ph.D. thesis, School of Engineering Sciences, Faculty of Engineering, Science & Mathematics, University of Southampton, UK (2009)
53. K.P. D'Epagnier, in *Ocean. 2006* (IEEE, 2006), pp. 1–7
54. A. Phillips, S. Turnock, M. Furlong, Comparisons of CFD simulations and in-service data for the self propelled performance of an Autonomous Underwater Vehicle (2008)
55. R.G. Coe, W.L. Neu, *Ocean 2012 MTS/IEEE Harnessing Power Ocean*, pp. 1–4 (2012)
56. L.K.P. D'Epagnier, H.L. Chung, M.J. Stanway, R.W. Kimball, in *Ocean Conference Records* (IEEE, 2007), pp. 1–8
57. A. Nemati Hayati, S.M. Hashemi, M. Shams, Ocean Eng. **59**, 152 (2013)
58. J.L.D. Dantas, E.A. de Barros, Appl. Ocean Res. **42**, 168 (2013)
59. M.E. Furlong, D. Paxton, P. Stevenson, M. Pebody, S.D. McPhail, J. Perrett, in *2012 IEEE/OES Autonomous Underwater Vehicles AUV 2012* (IEEE, 2012), pp. 1–7

60. F.-C. Chiu, J. Guo, J.-G. Chen, Y.-H. Lin, in *Proceedings 2002 International Symposium on Underwater Technology (Cat. No.02EX556)*, vol. 2002 (IEEE, 2002), pp. 172–177
61. J.E. Colgate, K.M. Lynch, IEEE, J. Ocean. Eng. **29**(3), 660 (2004)
62. M.F. Shaari, Z. Samad, C.J. Jun, H.A. B., A.M. Omar, in *2013 IEEE International Conference on Mechatronics and Automation* (IEEE, 2013), pp. 1594–1598
63. W. Khalil, G. Gallot, F. Boyer, IEEE Trans. Syst. Man Cybern. Part C (Appl. Rev.) **37**(6), 1259 (2007)
64. H.J. Kim, S.H. Song, S.H. Ahn, Smart Mater. Struct. **22**(1), 014007 (2013)
65. J. Yu, Y. Hu, R. Fan, L. Wang, J. Huo, Adv. Robot. **21**(3–4), 499 (2007)
66. J. Yu, M. Tan, S. Wang, E. Chen, IEEE Trans. Syst. Man, Cybern. Part B Cybern. **34**(4), 1798 (2004)
67. A. Willy, K. Low, in *2005 IEEE International Conference on Robotics and Biomimetics—ROBIO*, pp. 45–50 (2005)
68. A.N.A. Mazlan, A fully actuated tail propulsion system for a biomimetic autonomous underwater vehicle. Ph.D. thesis, University of Glasgow (2015)
69. M.S. Triantafyllou, A.H. Techet, F.S. Hover, IEEE, J. Ocean. Eng. **29**(3), 585 (2004)
70. P. Krishnamurthy, Khorrami, J. De Leeuw, M.E. Porter, K. Livingston, J.H. Long, *Proceedings of the American Control Conference ACC*, pp. 5224–5229 (2010)
71. A. Chemori, K. Kuusmik, T. Salumäe, M. Kruusmaa, in *IEEE International Conference on Robotics and Automation*, vol. 2016, pp. 4750–4755 (2016)
72. W.S. Chu, K.T. Lee, S.H. Song, M.W. Han, J.Y. Lee, H.S. Kim, M.S. Kim, Y.J. Park, K.J. Cho, S.H. Ahn, Review of biomimetic underwater robots using smart actuators. Int. J. Precis. Eng. Manuf. **13**(7), 1281 (2012)
73. L. Zhang, We, Y. Hu, D. Zhang, L. Wang, in *2007 IEEE/RSJ International Conference on Intelligent Robots and Systems*, vol. 2 (IEEE, 2007), pp. 3560–3565
74. J. Guo, Y.J. Joeng, Proc. Inst. Mech. Eng. Part M J. Eng. Marit. Environ. **218**(2), 93 (2004)
75. C. Zhou, Z. Cao, S. Wang, M. Tan, in *2006 IEEE/RSJ International Conference on Intelligent Robots and Systems* (IEEE, 2006), pp. 5406–5411
76. P. Babcock, J. Zinchuk, *Symposium on Autonomous Underwater Vehicle Technology*, vol. 4, 34 (1990)
77. A.B. Phillips, S.R. Turnock, M. Furlong, Proc. Inst. Mech. Eng. Part M J. Eng. Marit. Environ. **224**(4), 239 (2010)
78. N. Kato, J. Kojima, Y. Kato, S. Matumoto, K. Asakawa, in *Proceedings. 1998 IEEE International Conference on Robotics and Automation* (Cat. No.98CH36146), vol. 2 (IEEE, 1998), pp. 1045–1050
79. K. Alam, T. Ray, S.G. Anavatti, Proc. Inst. Mech. Eng. Part M J. Eng. Marit. Environ. **226**(3), 235 (2012)
80. K. Vasudev, R. Sharma, S. Bhattacharyya, in *Ocean. 2014 - TAIPEI* (IEEE, 2014), pp. 1–8
81. J. Sobieszczanski-Sobieski, in *Advanced Structural Optimisation*, vol. 25 (Springer Netherlands, 1995), pp. 483–496
82. C. McAllister, T. Simpson, P. Kurtz, M. Yukish, in *9th AIAA/ISSMO Symposium on Multidisciplinary Analysis and Optimization* (American Institute of Aeronautics and Astronautics, Reston, Virigina, 2002)
83. M. Martz, W. Neu, Ocean **2008**, 1–9 (2008)
84. G. Xia, C. Liu, X. Chen, in *Ocean. 2016 - Shanghai* (IEEE, 2016), pp. 1–6
85. T. Bäek, *Evolutionary Algorithms in Theory and Practice: Evolution Strategies, Evolutionary Programming, Genetic Algorithms* (Oxford University Press, 1996)
86. A.R. Yildiz, Eng. Appl. Artif. Intell. **26**(1), 327 (2013)
87. O. Chocron, P. Bidaud, *Proceedings of the 1997 IEEE/RSJ International Conference on Intelligent Robots and Systems. Innovations in Robotics. Real-World Applications IROS '97*, vol. 2 (February), p. 1111 (1997)
88. A. Bataller, J. Cabrera, M. Clavijo, J. Castillo, Mech. Mach. Theory **105**, 31 (2016)
89. O. Chocron, P. Bidaud, Conception Evolutionnaire de Systemes Robotiques. Ph.D. thesis, Université de Paris 06 (2000)

90. C. Darwin, Murray, London (1859)
91. D.E. Goldberg, *Genetic Algorithms in Search, Optimization, and Machine Learning* (Addison-Wesley Longman Publishing Co., Inc., 1989)
92. H.J. Lichtfuss, Evolution eines rohrkrümmers
93. I. Rechenberg, in *R. Aircr. Establ. Transl. No. 1122* (Farnborough, Hants., UK, 1965)
94. D.B. Fogel, *System Identification trough Simulated Evolution: A Machine Learning Approach* (Ginn Press, 1991)
95. M. Srinivas, L.M. Patnaik, Computer (Long. Beach. Calif). **27**(6), 17 (1994)
96. J.H. Holland, *Adaptation in Natural and Artificial Systems: An Introductory Analysis with Applications To Biology, Control, and Artificial Intelligence* (University of Michigan Press, 1975)
97. K.A. De Jong, An analysis of the behaviour of a class of genetic algorithms. Ph.D. thesis, University of Michigan (1975)
98. J.J. Grefenstette, I.E.E.E. Trans, Syst. Man Cybern. **16**(1), 122 (1986)
99. M. Molga, C. Smutnicki, *Test Functions for Optimization Needs* (2005)
100. S. Koos, J.B. Mouret, S. Doncieux, I.E.E.E. Trans, Evol. Comput. **17**(1), 122 (2013)
101. N. Jakobi, Minimal simulations for evolutionary robotics. Ph.D. thesis, University of Sussex (1998)
102. A. Thompson, P. Layzell, R.S. Zebulum, I.E.E.E. Trans, Evol. Comput. **3**(3), 167 (1999)
103. D. Floreano, J. Urzelai, Auton. Robots **11**(3), 311 (2001)
104. G. Hornby, S. Takamura, J. Yokono, O. Hanagata, T. Yamamoto, M. Fujita, in *Proceedings 2000 ICRA. Millennium Conference IEEE IEEE International Conference on Robotics and Automation Symposium. Proceedings* (Cat. No.00CH37065) vol. 3(April), p. 3040 (2000)
105. V. Zykov, J. Bongard, H. Lipson, Proc. Genet. ... (2004)
106. J.B. Pollack, H. Lipson, S. Ficici, P. Funes, G. Hornby, R.A. Watson, in *International Conference on Evolvable Systems*, vol. 1801 (Springer Berlin Heidelberg, 2000), pp. 175–186
107. S. Nolfi, D. Floreano, *Artificial Life. IV Proceedings of the 4th International Workshop on Artificial Life*, vol. 1997, p. 190 (1994)
108. J. Bongard, H. Lipson, ... Conf. Simul. ... pp. 57–62 (2004)
109. J. Bongard, V. Zykov, H. Lipson, Science **314**(5802), 1118 (2006)
110. J.C. Zagal, J. Ruiz-Del-Solar, J. Intell. Robot. Syst. Theory Appl. **50**(1), 19 (2007)

Erratum to: Fast Fourier-Based Block-Matching Algorithm for Sonar Tracks Registration in a Multiresolution Framework

Florian Nicolas, Andreas Arnold-Bos, Isabelle Quidu and Benoît Zerr

Erratum to: Chapter 1 in: L. Jaulin et al. (eds.), *Marine Robotics and Applications*, Ocean Engineering & Oceanography 10, https://doi.org/10.1007/978-3-319-70724-2_1

In the original version of the book, the following corrections have been incorporated in Chapter 1:

In page 3, the sentence "Indeed, while the former is obviously close to one, the latter can be neglected due to the INS heading drift which has experimentally been shown to be much slower than velocity and acceleration ones." has been replaced with "Indeed, we assume that these global parameters can be modeled through our block-matching formulation."

Table 1.1 has been updated with new corrections.

The updated online version of this chapter can be found at
https://doi.org/10.1007/978-3-319-70724-2_1

L. Jaulin et al. (eds.), *Marine Robotics and Applications*, Ocean Engineering & Oceanography 10, https://doi.org/10.1007/978-3-319-70724-2_10

Erratum to: Fast Fourier-Based Block-Matching Algorithm for Sonar Tracks Registration in a Multiresolution Framework

[illegible]

[illegible]

The manufacturer's authorised representative in the EU is Springer Nature Customer Service Centre GmbH, Europaplatz 3, 69115 Heidelberg, Germany. If you have any concerns regarding our products, please contact ProductSafety@springernature.com

Printed and bound by CPI Group (UK) Ltd, Croydon, CR0 4YY
05/10/2026
02228141-0019